Basic Asphalt Emulsion Manual

Printed in the USA
Fourth Edition
ISBN No. 978-1-934154-56-4

Library of Congress Control No. 2008927817

The primary purpose of this manual is to impart a basic understanding of asphalt emulsions to those who work with the product. Further, it is intended to be useful in choosing the emulsion that best fits a project's specific conditions. It should be most helpful in evaluating pavement systems for construction and maintenance. A thorough study of the manual should enable one to recommend where, when, and how emulsions should be used. It also should aid in the solving of problems that may arise on projects in which emulsions are used.

The manual is not written in such detail that one can use it to produce asphalt emulsions. Neither is it directed toward the specific features of one manufacturer's products. Rather, it explains the general characteristics of asphalt emulsions and their uses.

The Asphalt Institute and the Asphalt Emulsion Manufacturers Association have jointly published this Fourth Edition of MS-19, *Basic Asphalt Emulsion Manual*. This publication was revised by a team consisting of Keith Davidson (McAsphalt Industries Ltd.), Mark Ishee (Ergon Asphalt & Emulsions, Inc.), Arlis Kadrmas (SemMaterials, L.P.), and from the Asphalt Institute: Becky Boston, Mark Buncher, Wayne Jones, Mike Sonnenberg, and Dwight Walker.

ASPHALT INSTITUTE
Research Park Drive
2696 Research Park Drive
Lexington, Kentucky 45011-8480
(859) 288-4960
Fax (859) 288-4999

ASPHALT EMULSION MANUFACTURERS ASSOCIATION
#3 Church Circle, Suite 250
Annapolis, Maryland 21401
(410) 267-0023
Fax (410) 267-7546

TABLE OF CONTENTS

CHAPTER 1

INTRODUCTION

Virtually all types of asphalt used in the United States are products of the refining of crude petroleum. Asphalt is produced in a variety of types and grades ranging from hard and brittle solids to thin liquids. The asphalt used for paving is normally in the middle of these two extremes. Although paving asphalt is a semi-solid or solid at ambient temperatures, it can be readily liquefied by heating, by adding a petroleum solvent, or by emulsifying it in water.

In the production of hot mix asphalt (HMA), heat is used to liquefy the asphalt so it will coat the aggregate and remain workable during transport, laydown, and compaction. As the asphalt cools, it hardens and regains the binding properties that make it an effective paving material. When a petroleum solvent such as naphtha or kerosene is added to the base asphalt to make it fluid, the product is called cutback asphalt. As the solvent evaporates, the cutback cures and the asphalt's binding properties are restored.

When asphalt is milled into microscopic particles and dispersed in water with a chemical emulsifier, it becomes an asphalt emulsion. The tiny droplets of asphalt remain uniformly suspended until the emulsion is used for its intended purpose. In the emulsion state, the emulsifier molecules orient themselves in and around droplets of asphalt. The chemistry of the emulsifier/asphalt/water system determines the dispersion and the stability of the suspension. When emulsions are used in the field, the water evaporates into the atmosphere, and the chemical emulsifier is retained with the asphalt.

History of Asphalt Emulsion

Emulsions were first developed in the early 1900s. Emulsions came into general use in pavement applications in the 1920s. Their early use was in spray applications and as dust palliatives. The growth in the use of asphalt emulsions was relatively slow, limited by the type of emulsions available and a lack of knowledge as to how they should be used. Continuing development of new types and grades, coupled with improved construction equipment and practices, now gives a broad range of choices (see Table 1.1). Virtually any roadway requirement can be met with emulsions. Judicious selection and use can yield significant economic and environmental benefits. A slow but steady increase in the amount of emulsions used came about between 1930 and the mid-1950s. Following World War II, traffic loads and volumes increased so much that roadway designers began to curtail the use of asphalt emulsions. Instead, they specified hot mix asphalt requiring the use of asphalt. While the volume of asphalt cement used has increased greatly since 1953, the combined use of other asphalt products has remained almost constant. But there has been a steady rise in the volume of asphalt emulsions used.

◻ **Table 1.1: The Major Uses of Asphalt Emulsion**

Surface Treatments	Asphalt Recycling	Other Applications
• Scrub Seal	• Cold in-place	• Stabilization (soil and base)
• Chip Seal	• Full Depth	• Maintenance Patch
• Fog Seal	• Hot in-place	• Bond Coat (tack coat)
• Sand Seal	• Central Plant Mix	• Dust Palliative
• Slurry Seal		• Prime Coat
• Micro Surfacing		• Crack Fill
• Cape Seal		• Protective Coating

Several factors have contributed to interest in the use of asphalt emulsions:

- Asphalt emulsion does not require a petroleum solvent to make it liquid. Asphalt emulsions can also be used in most cases without additional heat. Both of these factors contribute to energy savings.
- Concerns about reducing atmospheric pollution. There are little or no hydrocarbon emissions from asphalt emulsions.
- The ability of certain types of asphalt emulsions to coat damp aggregate surfaces. This reduces the fuel requirements for heating and drying aggregates.
- Availability of a variety of emulsion types. New formulations and improved laboratory procedures have been developed to satisfy design and construction requirements.
- The ability to use the cold materials at remote sites.
- The applicability of emulsions for use in preventive maintenance to increase the service life of slightly distressed existing pavements.

Asphalt Emulsions in Pavement Preservation

Every road surface will eventually deteriorate over time due to aging, mechanical wear, and exposure to harsh climate conditions. Pavement preservation is a strategy of managing pavement condition to maximize the pavement's lifespan at the minimal cost. It can be applied to all types of roads, from low volume local roads to heavy interstates. This is achieved by careful planning and selection of the right protective treatment (application) at the optimal time. Many of these preservation treatment types utilize asphalt emulsions.

The decision to establish a pavement preservation program can be difficult for an agency because it often requires a change in the way an agency does business. Traditionally, agencies allocate funds to pavements in the worst condition. However, when a strong pavement preservation program is in place, agencies can address the worst pavements, while at the same time dedicating funds to keep good roads in good condition in order to delay the need for more costly treatments.

The benefits of a pavement preservation program can best be demonstrated through the use of a comprehensive pavement management system (described in the Asphalt Institute's MS-4, *The Asphalt Handbook*, Chapter 11, *Pavement Management*). The pavement management system should contain the following considerations in its analysis:

- A method of assessing pavement conditions that include the types of deterioration that normally trigger the selection of preventive maintenance and rehabilitation treatments.
- Pavement performance models that can differentiate the performance of a pavement section with and without different types of treatments.
- Treatment rules that can be used to determine when different types of treatments are considered feasible.
- Models that quantify the effects on pavement condition after different types of treatments have been applied. These models are used to forecast future conditions under various scenarios.

The appropriate amount of funds allocated to pavement preservation activities, versus rehabilitation and reconstruction, will depend on the current and projected condition of an agency's pavement network. By utilizing a pavement management system, an agency can evaluate different funding allocations and strategies to determine how they will meet the targeted condition levels (goals) established by the agency. These targeted conditions are generally a minimum condition level for each road classification (arterial, collector, local, etc.) above which the network should be maintained. The number of lane miles that do not meet these condition levels can be estimated along with the estimated cost of total repair needs. In addition, an effective pavement management system can identify the most appropriate maintenance and rehabilitation treatments for addressing the type and extent of pavement deterioration.

The demand for a well maintained and efficient highway network is higher than ever with the continued increase in traffic loads and volumes. Asphalt is essential to meeting this requirement, as 93 to 94 percent of all federally funded roads and highways are surfaced with asphalt (Annual FHWA Highway Statistics Report). The use of recycling technologies and pavement preservation techniques utilizing asphalt emulsions plays a significant role in meeting the challenges of increased traffic, shrinking agency budgets, and increased emphasis on environmental sustainability. A clear understanding of the "what," "how," and "why" for using asphalt emulsions can lead to their most effective and efficient use. This manual is directed towards these ends.

CHAPTER 2

CHEMISTRY

Many types of emulsion products are used in our daily lives. Mayonnaise, latex paint, and ice cream are some of the more common emulsions. In each case, certain mechanical and chemical processes are involved to combine two or more materials that will not mix under normal conditions. An entire scientific field is devoted to the study of emulsification. Just as an understanding of how an engine works is not necessary to operate an automobile, neither does one have to understand complex emulsion chemistry to successfully use an asphalt emulsion. The key is to select the right emulsion for the materials and application. Throughout this text, when the term "emulsion" is used, it is intended to mean "asphalt emulsion."

Composition of Asphalt Emulsions

An asphalt emulsion consists of three basic ingredients: asphalt, water, and an emulsifying agent. On some occasions, the emulsion may contain other additives, such as stabilizers, coating improvers, antistrips, or break control agents.

It is well known that water and asphalt will not mix, except under carefully controlled conditions using highly specialized equipment and chemical additives. The blending of asphalt and water is the same as an auto mechanic trying to wash grease from his hands with only water. Only with a detergent or soapy agent can grease be successfully removed. The soap particles surround the particles of grease, break the surface tension that holds them, and allow them to be washed away.

Some of the same physical and chemical principles apply in the formulation, production, and use of asphalt emulsions. The object is to make a stable dispersion of the asphalt cement in water—stable enough for pumping, prolonged storage, and mixing. Furthermore, the emulsion should "break" quickly after contact with aggregate in a mixer or after spraying on the roadbed—breaking is the separation of the water from the asphalt. Upon curing, the residual asphalt retains all of the adhesion, durability, and water-resistance of the asphalt cement from which it was produced.

Asphalt Emulsion Classification

Asphalt emulsions are classified into three categories: anionic, cationic, and nonionic. In practice, the first two types are more widely used in roadway construction and maintenance. Nonionics may become more important as emulsion technology advances. The anionic and cationic classes refer to the electrical charges surrounding the asphalt particles. This identification system stems from a basic law of physics—like charges repel one another and unlike charges attract.

When two poles (an anode and a cathode) are immersed in a liquid and an electric current is passed through, the anode becomes positively charged and the cathode becomes negatively charged. If a current is passed through an emulsion containing negatively charged particles of asphalt, they will migrate to the anode. Hence, the emulsion is referred to as anionic. Conversely, positively charged asphalt particles will move to the cathode and the emulsion is known as cationic. With nonionic emulsions, the asphalt particles are neutral and do not migrate to either pole.

Emulsions are further classified on the basis of how quickly the asphalt droplets will coalesce; i.e., revert to asphalt cement. The terms RS, MS, SS, and QS have been adopted to simplify and standardize this classification. They are relative terms only and mean rapid-setting, medium-setting, slow-setting, and quick-setting. The tendency to coalesce is closely related to the speed with which an emulsion will become unstable and break after contacting the surface of an aggregate. An RS emulsion has little or no ability to mix with an aggregate, an MS emulsion is expected to mix with coarse but not fine aggregate, and SS and QS emulsions are designed to mix with fine aggregate, with the QS expected to break more quickly than the SS.

Emulsions are further identified by a series of numbers and letters related to viscosity of the emulsions and hardness of the base asphalt cements. The letter "C" in front of the emulsion type denotes cationic. The absence of the "C" denotes anionic in American Society for Testing and Materials (ASTM) and American Association of State Highway and Transportation Officials (AASHTO) specifications. For example, RS-1 is anionic and CRS-1 is cationic.

The numbers in the classification indicate the relative viscosity of the emulsion. For example, an MS-2 is more viscous than an MS-1. The "h" that follows certain grades simply means that harder base asphalt is used.

The "HF" preceding some of the anionic grades indicates high-float, as measured by the float test. High-float emulsions have a gel quality, imparted by the addition of certain chemicals that permit a thicker asphalt film on the aggregate particles and prevent drain off of asphalt from the aggregate. These grades are used primarily for cold and hot plant mixes, chip seals, and road mixes.

ASTM and AASHTO have developed standard specifications for the grades of emulsions listed in Table 2.1.

☐ Table 2.1: ASTM and AASHTO Standardized Emulsion Grades

Asphalt Emulsion (ASTM D 977, AASHTO M 140)	Cationic Asphalt Emulsion (ASTM D 2397, AASHTO M 208)	Polymer-Modified Cationic Emulsified Asphalt (AASHTO M 316)
RS-1	CRS-1	—
RS-2	CRS-2	CRS-2P, CRS-2L
HFRS-2	—	—
MS-1	—	—
MS-2	CMS-2	—
MS-2h	CMS-2h	—
HFMS-1	—	—
HFMS-2	—	—
HFMS-2h	—	—
HFMS-2s	—	—
SS-1	CSS-1	—
SS-1h	CSS-1h	—
QS-1h	CQS-1h	—

Most producers do not stock all grades of emulsion. As well, many users have their own specifications that do not follow ASTM or AASHTO guidelines for naming emulsions. Communication and planning between user and producer will help facilitate service and supply of a given grade.

Micro surfacing uses an emulsion often referred to as CSS-1hP. As with quick set emulsions, micro surfacing emulsions are required to meet ASTM and AASHTO CSS-1h requirements with the exception of the cement mixing test. In addition, a minimum polymer content normally is specified as 3 percent of solids based on the weight of the asphalt in the emulsion. This addition enhances the high temperature performance of the asphalt and permits application of micro surfacing in wheel ruts and other areas where multiple stone depths are required.

The expanding use of polymer-modified asphalts has contributed a whole new family of emulsion grades. Adding one letter (usually P, S, or L) to the end of the grade (e.g., HFRS-2P) normally designates modified emulsions.

Cationic emulsion specifications (ASTM D 2397, AASHTO M 208) permit solvent in some grades but restrict the amount. Some user agencies specify an additional cationic sand-mixing grade designated CMS-2s that contains more solvent than other cationic grades.

The general uses of asphalt emulsion are given in Table 5.1 and are discussed in detail later in this publication.

Variables Affecting Emulsion Quality

There are many factors that affect the production, storage, and performance of an asphalt emulsion. It would be hard to single out any one as being the most significant. Variables having a significant effect include:

- Chemical properties of the base asphalt cement
- Hardness and quantity of the base asphalt cement
- Asphalt particle size in the emulsion
- Type and concentration of the emulsifying agent
- Manufacturing conditions such as temperature, pressure, and shear
- Ionic charge on the emulsion particles
- Order of addition of the ingredients
- Type of equipment used in manufacturing, storage, and application of the emulsion
- Properties of the emulsifying agent
- Addition of chemical modifiers or polymers
- Water quality (hardness)

These factors can be varied to suit the available aggregates or construction conditions. It is always advisable to consult the emulsion supplier with respect to a particular asphalt-aggregate combination, as there are few rules that apply under all conditions.

Emulsion Ingredients

An examination of the three main constituents—asphalt, water, and emulsifier—is essential to an understanding of why asphalt emulsions work as they do.

Asphalt

Asphalt cement is the basic ingredient of asphalt emulsion and, in most cases, it makes up from 50 to 75 percent of the emulsion. Asphalt chemistry is a complex subject, and there is no need to examine all the properties of asphalt cement. Some properties of the asphalt cement do significantly affect the finished emulsion. There is not an exact correlation, however, between the properties and the ease with which the asphalt can be emulsified. Although hardness of base asphalt cements may vary, most emulsions are made with asphalts in the 40–250 penetration range. On occasion, climatic conditions may require harder or softer base asphalt. In any case, chemical compatibility of the emulsifying agent with the asphalt cement is essential for production of a stable emulsion.

The principal source of asphalt is the refining of crude petroleum. Asphalt is primarily composed of large hydrocarbon molecules, and its chemical composition is diverse. The colloidal makeup of the asphalt depends on the chemical nature and percentage of the hydrocarbon molecules and their relationship to each other. The varying chemical and physical characteristics of asphalt, therefore, are primarily due to inherent variations in crude oil sources and refining practices. The properties of the asphalt cement will have an effect on the performance of the residual asphalt on the road.

The complex interaction of the different molecules makes it almost impossible to predict accurately the behavior of an asphalt to be emulsified. For this reason, quality control is maintained on emulsion production. Each emulsion manufacturer has its own formulations and production techniques. They have been developed to achieve optimum results with the asphalt cement and emulsifying chemicals that are used.

Water

The second ingredient in an asphalt emulsion is water. Its contribution to the desired properties of the finished product cannot be minimized. Water may contain minerals or other matter that affect the production of stable asphalt emulsions. Accordingly, water considered suitable for drinking might not be suitable for asphalt emulsions.

Water found in nature may be unsuitable because of impurities, either in solution or colloidal suspension. Of particular concern is the presence of calcium and magnesium ions. These ions benefit the formation of a stable cationic emulsion. In fact calcium chloride is often added to cationic emulsions to enhance storage stability. These same ions, however, can be harmful in anionic emulsions. This is because water-insoluble calcium and magnesium salts are formed in the reaction with water soluble sodium and potassium salts normally used as emulsifiers. In a like manner, carbonate and bicarbonate anions can help stabilize an anionic emulsion because of their buffering effect, but they can destabilize cationic emulsions by reacting with water soluble amine hydrochloride emulsifiers.

Water containing particulate matter should not be used in emulsion production. It can be especially harmful in cationic emulsions. The usually negatively charged particles quickly absorb the cationic emulsifying agents, destabilizing the emulsion. The use of impure water may result in an imbalance of the emulsion components that can adversely affect performance or cause premature breaking.

Emulsifying Agents

Asphalt emulsion properties greatly depend on the chemical used as the emulsifier. The emulsifier is a surface-active agent, or surfactant. The emulsifier keeps the asphalt droplets in stable suspension and controls the breaking time. It is also the determining factor in the classification of the emulsion as anionic, cationic, or nonionic.

In the early days of asphalt emulsion production, materials such as ox-blood, clays, and soaps were used as emulsifying agents. As emulsion demand increased, more efficient emulsifying agents were found. Many chemical emulsifiers are now commercially available.

The most common anionic emulsifiers are fatty acids, which are wood-product derivatives such as tall oils, rosins, and lignins. Anionic emulsifiers are saponified (turned into soap) by reacting with sodium hydroxide or potassium hydroxide.

Most cationic emulsifiers are fatty amines (e.g., diamines, imidazolines, and amidoamines). The amines are converted into soap by reacting with acid, usually hydrochloric. Another type of emulsifying agent, fatty quaternary ammonium salts, is used to produce cationic emulsions. They are water soluble salts and do not require the addition of acid. They are stable, effective cationic emulsifiers.

By broad definition, surfactants are water soluble substances whose presence in solution markedly changes the properties of the solvent and the surfaces they contact. They are categorized by the way they dissociate or ionize in water. Structurally, they possess a molecular balance of a long lipophilic (oil-loving), hydrocarbon tail and a polar, hydrophilic (water-loving) head. Surfactants are adsorbed at the interface between liquids and gases or liquid and solid phases. They tend to concentrate at the interface such that the hydrophilic groups orient themselves towards the more polar phase, and the lipophilic groups towards the less polar phase. The surfactant molecule or ion acts as a bridge between two phases.

Basically, there are three types of surfactants that are classified according to their dissociation characteristics in water:

1. **Anionic Surfactants** – Where the electrovalent and polar hydrocarbon group is part of the negatively charged ion, when the compound ionizes:

ANIONIC

$CH_3(CH_2)_n \, COO^- Na^+$

2. **Nonionic Surfactants** – Where the hydrophilic group is covalent and polar, and which dissolves without ionization:

NONIONIC

$CH_3(CH_2)_n \, COO \, (CH_2CH_2O)_x H$

3. **Cationic Surfactants** – Where the electrovalent and polar hydrocarbon group is part of the positively charged ion when the compound ionizes:

CATIONIC

$CH_3(CH_2)_n \, NH_3^+ Cl^-$

The emulsifier is the single most important component in any asphalt emulsion formulation. To be an effective emulsifier for asphalt, a surfactant must be water soluble and possess a proper balance between the hydrophilic and lipophilic properties. The emulsifier, used in combination with an acceptable asphalt, good quality water, and adequate mechanical input, is the major factor in emulsification, emulsion stability, and field performance.

Each manufacturer has its own procedure for using surfactants in asphalt emulsion production. In most cases, the surfactant is combined with the water before introduction into the colloid mill.

Producing the Emulsion

Emulsifying Equipment

The basic equipment to prepare an emulsion includes a high-speed, high-shear mechanical device (usually a colloid mill) to divide the asphalt into tiny droplets. A schematic diagram of a typical asphalt emulsion manufacturing plant is shown in Figure 2.1. Also needed are an emulsifier solution tank, heated asphalt tank, pumps, and flow-metering gauges.

The colloid mill has a high-speed rotor that revolves at 1,000–6,000 revolutions per minute (17–100 Hertz) with mill clearance settings in the range of about 0.01 to 0.02 inches (0.25 to 0.50 millimeters). Typically, asphalt emulsions have droplet sizes smaller than the diameter of a human hair, or about 0.00004 to 0.0004 inches (0.001 to 0.010 millimeters). Particle size analyzers are commonly used to characterize emulsion quality. Asphalt droplet sizes depend upon the mechanical energy density imparted by the mill.

Separate pumps are used to deliver asphalt and the emulsifier solution into the colloid mill. Because the emulsifier solution can be corrosive, it may be necessary to use equipment made of corrosion resistant materials.

The Emulsifying Process

Prior to the emulsification process, asphalt and the water are heated separately to the optimum temperature. The asphalt is fed into the colloid mill where it is divided into tiny droplets. At the same time, water containing the emulsifying agent is fed into the colloid mill. These temperatures vary and depend upon the properties of the asphalt

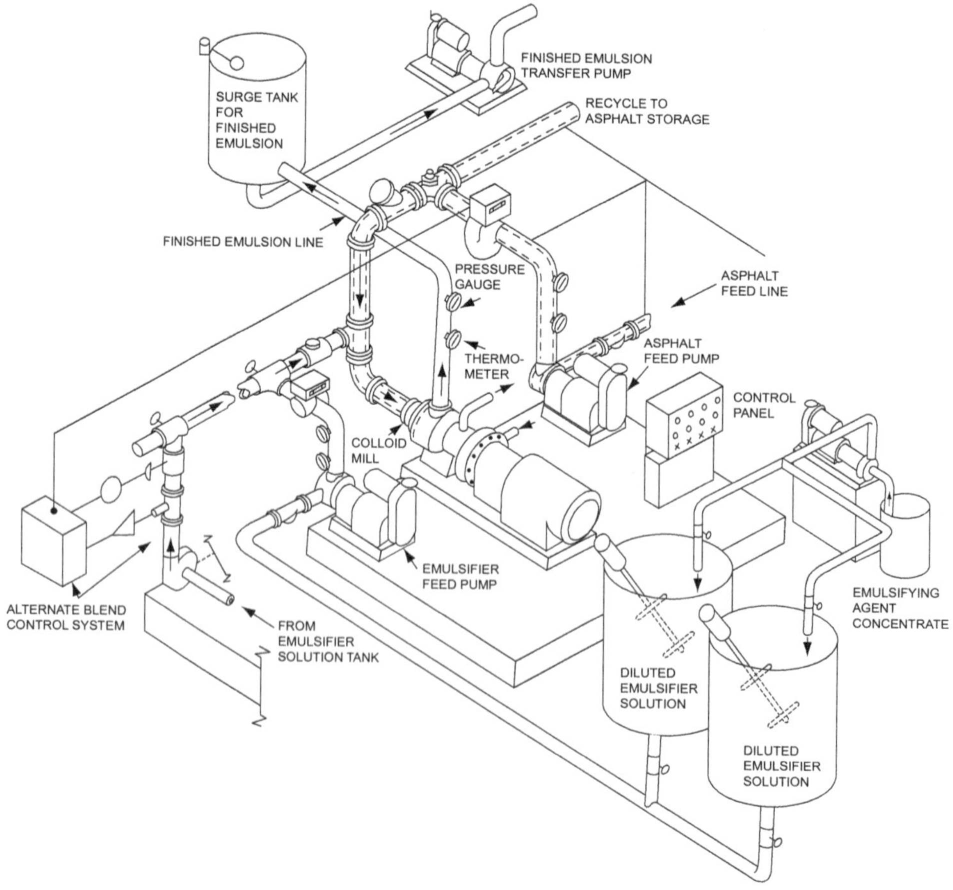

☐ **Figure 2.1** Diagram of an Asphalt Emulsion Manufacturing Plant

and the compatibility between the asphalt and the emulsifying agent. Unless a heat exchanger is used to cool the emulsion, the temperature of the emulsion leaving the mill must be below the boiling point of water. The emulsion is then usually pumped into bulk storage tanks. These tanks may be equipped with mechanical agitation to keep the emulsion uniformly blended.

The method of adding the emulsifier to the water varies according to the manufacturer's procedure. Some emulsifiers, such as amines, must be mixed and reacted with an acid to be water soluble. Others, such as fatty acids, must be mixed and reacted with an alkali to be water soluble. Emulsifier mixing is typically done in a batch mixing tank. The emulsifier is introduced into warm water containing acid or alkali and agitated until completely dissolved.

Asphalt and emulsifier solution must be proportioned accurately. This is normally done with flow meters. Proportioning can also be accomplished by monitoring the temperatures of the asphalt and emulsifier solution entering the mill, and the discharge temperature.

Asphalt particle size is a vital factor in making a stable emulsion. A microscopic photograph of a typical emulsion (Figure 2.2) reveals these average particle sizes:

- Smaller than 0.001 millimeter (1 micron)   20 percent
- 0.001–0.005 millimeter (1–5 microns)   57 percent
- 0.005–0.010 millimeter (5–10 microns)   23 percent

These microscopic-sized asphalt droplets are dispersed in water in the presence of the chemical surfactant. The surfactant causes a change in the surface tension at the contact area between the asphalt droplets and the surrounding water, and this allows the asphalt to remain in a suspended state. The asphalt particles, all having a similar electrical charge, repel each other, which aids in keeping them suspended.

Breaking and Curing

Breaking

If the asphalt emulsion is to perform its ultimate function as a binder, the water must separate from the asphalt phase and evaporate. This separation is called "breaking." Emulsions are formulated to break according to their intended use. The two mechanisms by which asphalt emulsions break are chemically and evaporative.

For the slow-setting grades, the breaking mechanism is mainly evaporation. For the medium-setting and rapid-setting grades, the breaking mechanism is mainly chemical. A rapid set emulsion will have a shorter breaking time, whereas a medium or slow set material may take considerably longer. For dense mixtures, more time is needed to allow for mixing and placement. Therefore, emulsions used for mixtures are formulated for delayed breaking.

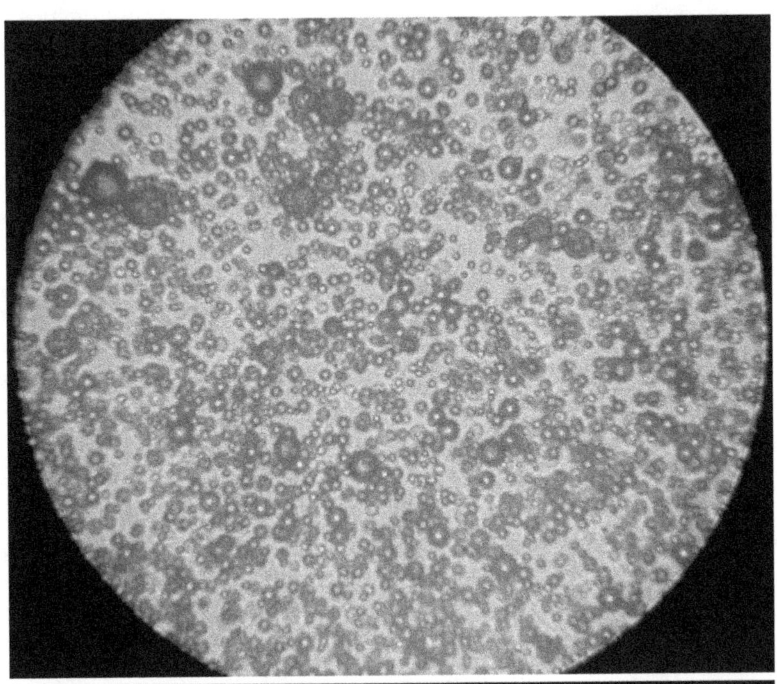

○ **Figure 2.2**   Relative Sizes and Distribution of Asphalt Particles in an Emulsion

The specific type and concentration of emulsifying agent primarily controls the rate of breaking. However other factors, discussed below, also play an important role in breaking the emulsion. In order to achieve optimum results, it is necessary to control all of these factors to meet the specific requirements of the field use of the asphalt emulsion. Contact the supplier for more information regarding the optimum use of emulsions.

Curing

Curing involves the development of the mechanical properties of the asphalt. For this to happen, the water must completely evaporate and the asphalt emulsion particles have to coalesce (come together) and bond to the intended surface. The water is removed by evaporation and absorption. When medium-setting and slow-setting grades are used for paving mixes, the use of slightly damp aggregates facilitates the mixing and coating process. The development of strength in the slow-setting grades depends mainly on evaporation and absorption.

Some asphalt emulsions may contain petroleum solvents to aid in the mixing and coating process. The type and quantity of petroleum solvent affect the speed of the curing process.

Factors Affecting Breaking and Curing

Some of the factors affecting breaking and curing rates of asphalt emulsions include:

- Water Absorption—A rough-textured, porous surface speeds the setting time by absorbing water from the emulsion.
- Aggregate Moisture Content—While wet aggregate may facilitate coating, it tends to slow the curing process by increasing the amount of time needed for evaporation.
- Weather Conditions—Temperature, humidity, and wind velocity all have a bearing on water evaporation rate, emulsifier migration, and water release characteristics. While breaking usually occurs more quickly at warmer temperatures, that is not always the case. Hot weather can cause skin formation on chip seals, trapping water and delaying curing. Some chemical formulations have also recently been developed to break rapidly at cool temperatures.
- Mechanical Forces—Roller pressure and, to a limited extent, slow moving traffic forces the water from the mix and helps attain mix cohesion, cure, and stability.
- Surface Area—Greater aggregate surface area, particularly excessive fines or dirty aggregate, accelerates the breaking of the emulsion.
- Surface Chemistry—Intensity of the aggregate surface charge, in combination with the intensity of emulsifier charge, can impact setting rate, particularly for cationic emulsions. Calcium and magnesium ions on the aggregate surface can react with and destabilize certain anionic emulsifiers, accelerating setting.
- Emulsion and Aggregate Temperature—Breaking is accelerated when emulsion temperatures are high. This is particularly evident in micro surfacing.
- Type and Amount of Emulsifier—The surfactant determines the breaking characteristics of the emulsion.

These factors must be considered in determining working time after the emulsion has been sprayed or mixed with the aggregate in the field. The emulsion supplier is the best source of information.

CHAPTER 3

STORING, HANDLING, AND SAMPLING

Through a long history of successful use of asphalt emulsions, these handling, storing and sampling procedures have been established. These general guidelines should be followed. Questions about the storing, handling, or sampling of asphalt emulsions should be referred to the emulsion supplier.

Storing Asphalt Emulsions

Asphalt emulsion is a dispersion of fine droplets of asphalt in water and has special storage requirements. Storage tanks should be insulated for protection from freezing and efficient use of heat. A skin of asphalt can form on the surface of emulsions when exposed to air. Therefore, it is best to use tall, vertical tanks that expose the least amount of surface area to the air. Most fixed storage tanks are vertical, but horizontal tanks are often used for short-term field storage. Skin formation can be reduced by keeping horizontal tanks full to minimize the area exposed to air.

Side entering propellers located about 3 feet (1 meter) from the tank bottom may be used to prevent surface skin formation. These propellers should be large in diameter and turned slowly (approximately 60 RPM) to gently circulate the material. Propellers should only be utilized when there is sufficient emulsion for proper mixing.

When storing emulsified asphalts:

DO	store between 50°F (10°C) and 185°F (85°C), depending on the intended use and specific product.
DO	store at the temperature specified for the particular grade. Table 3.1 shows the normal storage temperature ranges.
DO NOT	permit the asphalt emulsion to be heated above 185°F (85°C). Elevated temperatures evaporate the water, changing the characteristics of the asphalt emulsion.
DO NOT	let the emulsion freeze. This breaks the emulsion, separating the asphalt from the water. The result will be two layers in the tank, neither of which will be suited for the intended use, and the tank will be difficult to empty.
DO NOT	allow the temperature of the heating surface to exceed 212°F (100°C). This will cause premature breakdown of the emulsion on the heating surface.
DO NOT	use forced air to agitate the emulsion. It may cause the emulsion to break.
DO NOT	allow excessive agitation by mixing or pumping.

 Table 3.1: Storage Temperatures for Asphalt Emulsions

Grade	Temperature °F (°C)	
	Minimum	Maximum
CQS-1h, QS-1h, Micro Surfacing Emulsion	50° (10°)	125° (50°)
RS-2, CRS-1, CRS-2, HFRS-2, CMS-2, CMS-2h, MS-2, MS-2h, HFMS-2, HFMS-2h	125° (50°)	185° (85°)
RS-1, SS-1, SS-1h, CSS-1, CSS-1h, MS-1	50° (10°)	140° (60°)

Handling Asphalt Emulsions

DO provide adequate ventilation. Avoid exposure to fumes, vapors, and mist.

DO obtain a copy of the supplier's material safety data sheet (MSDS). Read the MSDS carefully and follow it.

DO agitate gently when heating asphalt emulsion to eliminate or reduce skin formation.

DO protect pumps, valves, and lines from freezing in winter. Drain pumps and service according to the manufacturer's recommendations.

DO clear out lines and leave drain plugs open when not in service.

DO use pumps with proper clearances for handling emulsified asphalt. Tightly fitting pumps can bind and seize.

DO warm the pump to about 150°F (65°C) to facilitate start up.

DO (when diluting asphalt emulsion) check the compatibility of the water with the emulsion by testing a small quantity.

DO if possible, use warm water for diluting, and always add the water slowly to the emulsion (not the emulsion to the water).

DO avoid repeated pumping and recirculating, as the viscosity may drop and air may become entrained, causing the emulsion to be unstable.

DO place inlet pipes and return lines at the bottom of tanks to prevent foaming.

DO pump from the bottom of the tank to minimize contamination from skin fomation that may have formed.

DO haul emulsion in truck transports with baffle plates to prevent sloshing.

DO agitate emulsions slowly that have been in prolonged storage.

DO remember that emulsions with the same grade designation can be very different chemically and in performance. Consult the emulsion suppliers for compatibility before mixing different emulsions of the same grade.

DO NOT mix different classes, types, and grades of emulsified asphalt in storage tanks, transports, and distributors.

DO NOT load asphalt emulsion into storage tanks, tank cars, tank transports, or distributors containing remains of incompatible materials. See Tables 3.2, 3.3, and 3.4.

DO NOT apply severe heat to pump packing glands or pump casings.

DO NOT dilute rapid-setting grades of asphalt emulsion with water. Medium- and slow-setting grades may be diluted, but always add water slowly to the asphalt emulsion. Never add the asphalt emulsion to water when diluting.

DO NOT subject asphalt emulsion or air above it to an open flame, heat, or strong oxidants.

Contamination of materials can occur. Table 3.2 through Table 3.4 address the possibilities of contamination and give precautionary guidelines.

Table 3.2 provides guidelines for loading emulsions at the plant.

☐ Table 3.2: Condition of Transport Containers Before Loading

Product to be Loaded	Last Product in Tank					
	Asphalt Cement	Cutback Asphalt and Residual Fuel Oils	Cationic Emulsion	Anionic Emulsion	Crude Petroleum	Any Product Not Listed
Cationic Emulsion	Empty*	Empty to no measurable quantity	OK to load	Empty to no measurable quantity	Empty to no measurable quantity	Tank must be cleaned
Anionic Emulsion	Empty*	Empty to no measurable quantity	Empty to no measurable quantity	OK to load	Empty to no measurable quantity	Tank must be cleaned

*NOTE: Any material remaining may produce dangerous conditions.

Table 3.3 lists some of the possible causes of contamination during transportation and provides precautions.

☐ Table 3.3: Causes of Contamination During Transport

Possible Causes	Precautions
(a) Previous load not compatible with emulsion being loaded.	Examine the log of loads hauled or check with the supplier to determine if previous material hauled is detrimental. If it is, make sure vehicle tanks, unloading lines, and pump are properly cleaned and drained before being presented for loading. Provide a ramp at the unloading point at the plant to ensure complete drainage of vehicle tank while material is still fluid.
(b) Remains of diesel oil or solvents used for cleaning and flushing of tanks, lines, and pump.	When this is necessary, make sure all solvents are completely drained.
(c) Flushing of solvents into receiving storage tank or equipment tanks.	Do not allow even small amounts to flush into storage tank; entire contents may be contaminated.

Table 3.4 lists some of the possible causes of contamination and suggested precautions at the job site.

☐ Table 3.4: Causes of Contamination at Job Site

Possible Causes	Precautions
(a) Previous material left over in tank.	Any material allowed to remain must not cause the emulsion to become out of specification. If in doubt, refer to Table 3.2 and check with the emulsion supplier.
(b) Solvents used to flush transport.	Caution driver about flushing cleaning materials into storage tank.
(c) Flushing of lines and pump between storage tank and mixing plant with solvents and then allowing this material to return to the tank.	If necessary to flush lines and pump, suggest providing bypass valves and lines to prevent solvents from returning to the tank. A better solution is to provide insulated, heated lines and pump, thereby eliminating the necessity of flushing.
(d) Cleaning of distributor tank, pump, spray bar, and nozzles with solvents.	Be sure all possible cleaning material is drained or removed prior to loading.
(e) Dilutions from hot oil heating systems.	Check reservoir on hot oil heating system. If oil level is low, or oil has been added, check system for leakage into the emulsion.

Sampling Asphalt Emulsions

The purpose of any sampling method is to obtain samples that will show the true nature and condition of the material. The general procedure is described below. The standard procedure is further detailed in "Standard Methods of Sampling Bituminous Materials," ASTM D 140 or AASHTO T 40.

Containers for sampling asphalt emulsion shall be wide-mouth jars or bottles made of plastic, wide-mouth plastic-lined cans with lined screw caps, or plastic-lined triple-seal friction-top cans. The size of samples shall correspond to the required sample containers, which is generally 1 gallon (4 liters).

Whenever practical, the asphalt emulsion shall be sampled at the point of manufacture or storage. If that is not practical, samples shall be taken from the shipment immediately upon delivery. Three samples of the asphalt emulsion shall be taken. The samples shall be sent as soon as possible to the laboratory for testing.

Sampling Precautions

- Sample containers shall be new. They shall not be washed or rinsed. If they contain evidence of solder flux, or if they are not clean and dry, they shall be discarded. Top and container shall fit together tightly.
- Care shall be taken to prevent the samples from becoming contaminated. (See Table 3.5.) The sample container shall not be submerged in solvent, nor shall it be wiped with a solvent saturated cloth. Any residual material on the outside of the container shall be wiped with a clean, dry cloth immediately after the container is sealed and removed from the sampling device.
- The sample shall not be transferred to another container.
- The filled sample container shall be tightly and positively sealed immediately after the sample is taken.

Test results are greatly dependent upon proper sampling techniques. Extra care is required by the sampler to obtain samples that are truly representative of the material being sampled and will do much to eliminate the possibility of erroneous test results by reason of improper sampling. Make sure samples are taken only by those authorized persons who are trained in sampling procedures. Table 3.5 lists possible causes and precautions concerning non-representative samples.

☐ **Table 3.5: Causes of Non-Representative Samples**

Possible Causes	Precautions
(a) Contaminated sampling device (commonly called a "sample thief").	If sampling device (described in ASTM D 140 or AASHTO T 40) is cleaned with diesel oil or solvent, make sure that it is thoroughly drained and then rinsed out several times with emulsion being sampled prior to taking sample.
(b) Samples are taken with sampling device from top of tank where, under certain conditions, contaminants can collect on the surface.	In taking samples from the top of a tank, lower the sampling device at least one foot below the emulsion surface.
(c) Contaminated sample container.	Use only new clean containers. Never wash or rinse a sample container with solvent. Wide-mouthed plastic jars or bottles or plastic-lined cans should be used.
(d) Sample contaminated after taking.	DO NOT submerge container in solvent or wipe the outside of the container with a solvent-saturated rag. If necessary to clean spilled emulsion from the outside of container, use a clean, dry rag. Make sure the container lid is tightly sealed prior to storage or shipment. Ship to testing laboratory promptly.
(e) Samples are taken from spigot in lines between storage tank and mixing plant.	DO NOT take a sample while the hauling vehicle is pumping into the storage tank. DO NOT take a sample without allowing enough time for circulation and thorough mixing of emulsion. DO stop the pump prior to taking a sample from an inline spigot. DO drain sufficient material through spigot prior to taking a sample to ensure removal of any material lodged in spigot. DO take a sample slowly during circulation to be more representative of the emulsion being used.
(f) Samples taken from unloading line of hauling vehicle.	Drain sufficient emulsion through spigot prior to taking a sample to ensure removal of any material lodged there. Sample should be taken after one-third and not more than two-thirds of the load have been removed. Take the sample slowly to be sure it is representative of the emulsion being used.

Protection and Preservation of Samples

- Immediately after filling, sealing, and cleaning, the sample containers shall be properly marked for identification with a permanent marker on the container itself, not on the lid.
- Samples of emulsions shall be packaged, labeled, and protected from freezing during shipment.
- All samples should be packaged and shipped to the laboratory the same day they are taken. The containers should be tightly sealed and packed in protective material to reduce the probability of damage during shipment.
- Emulsion samples should be tested within two weeks from date of sampling.
- Each sample should be identified with this information:
  - Shipper's name and bill of lading or loading slip number
  - Date sampled
  - Sampler's name
  - Product grade
  - Project identification
  - Other important information as necessary

Safety Precautions

Safety precautions are mandatory at all times when sampling emulsions. These safety precautions include, but are not limited to the following:

- Gloves shall be worn and sleeves shall be rolled down and fastened over the gloves at the wrist while sampling and while sealing containers.
- Face shields should be worn while sampling.
- There shall be no smoking while sampling.
- Avoid prolonged breathing of fumes, vapors, and mists.
- During sealing and wiping, the container shall be placed on a firm level surface to prevent splashing, dropping, or spilling the material.

CHAPTER 4

TESTING

Proper interpretation of laboratory test results can greatly aid in determining the traits of an asphalt emulsion. Some of these tests are designed to measure performance qualities. Others deal with composition, consistency, and stability of the material. Laboratory tests are normally performed for these purposes:

- To provide data for specification requirements.
- To assess the quality and uniformity of the product during manufacturing and use.
- To predict the handling, storage, and field performance properties of the material.

A review of emulsion specifications used across the United States reveals a wide variety of requirements. Many are directly related to the emulsions produced by specific manufacturers. Because it is impractical to discuss the multitude of requirements and test methods, this chapter deals primarily with the methods in ASTM D 244 and AASHTO T 59. These standards describe the majority of test methods used to characterize emulsions. ASTM is in the process of creating individual standards for some of these test methods. Also included in this chapter are a few non-ASTM tests that are often used.

As advances are made in asphalt emulsion technology, corresponding advances in testing will evolve. This chapter will first consider the tests that apply to asphalt emulsions. It then describes the tests run on the asphalt emulsion residue after the water has been removed by distillation or evaporation.

Asphalt Emulsion

Proper sample handling is important to achieve valid test results. Follow the test method. Care should be taken to adjust the temperature of the emulsion sample in accordance with the test method. Samples should be stirred, not shaken, to ensure homogeneity.

Particle Charge (ASTM D 244)

The particle charge test is used to identify cationic asphalt emulsions. The test is conducted by heating a sample emulsion to 122°F (50°C) and pouring it into a 250-milliliter beaker. A positive electrode (anode) and a negative electrode (cathode) are then immersed in the sample and connected to a controlled direct current electrical source providing a current of at least 8 milliamperes (see Figure 4.1). After 30 minutes, or after the current has dropped to 2 milliamperes, the two electrodes are examined. An asphalt deposit on the cathode that is clearly discernible (compared with the anode) indicates that the sample is a cationic asphalt emulsion. **Note:** This test may not produce conclusive results for CSS grades. If the test is inconclusive, then a mixing test using silica sand is recommended (as described in ASTM D 244).

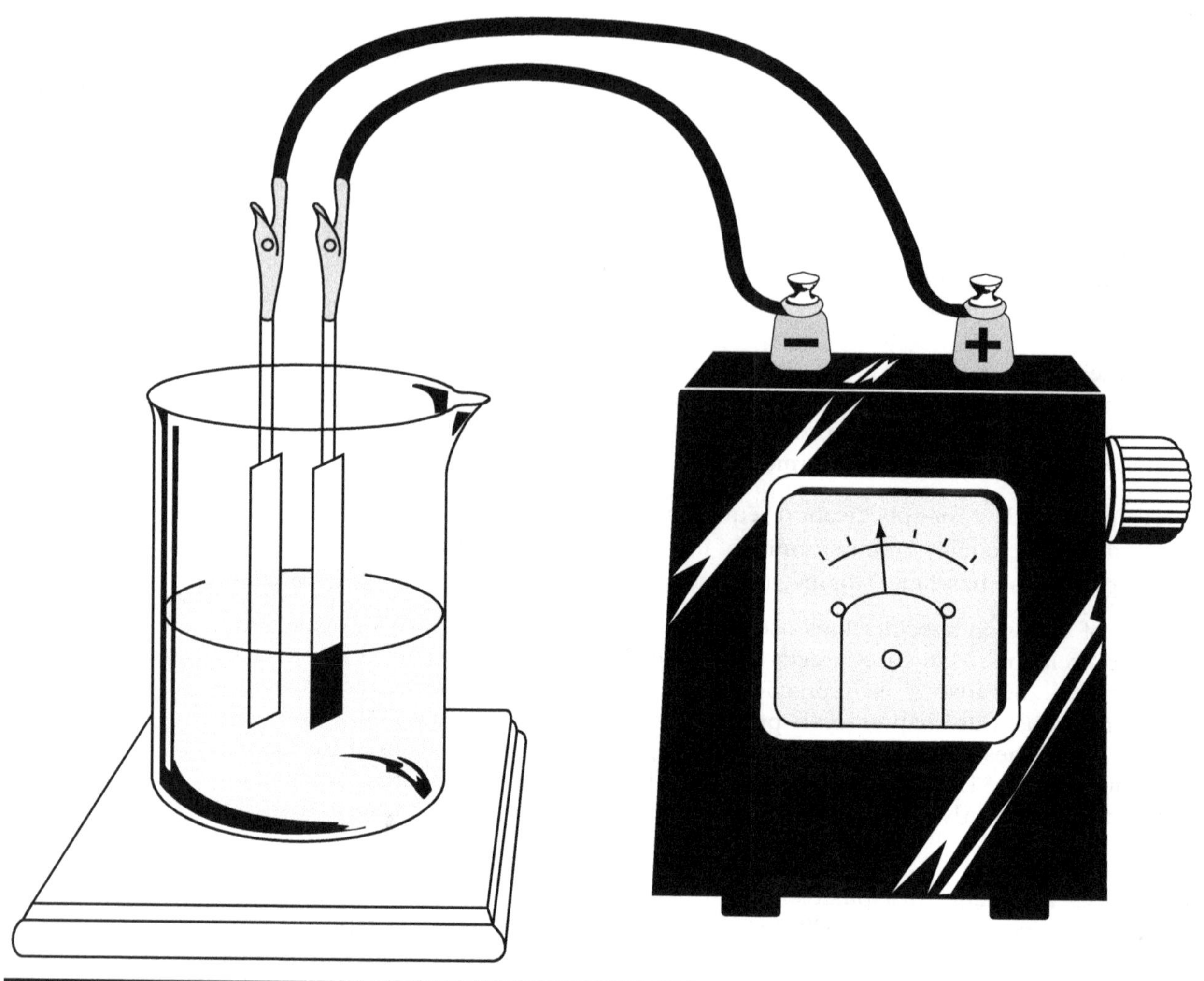

Figure 4.1 Particle Charge Test

Viscosity (ASTM D 244)

The Saybolt Furol viscosity test is used to measure the consistency (rate of flow) of asphalt emulsions. As a matter of testing convenience, and also to achieve suitable accuracy, the test is performed at one of two temperatures, 77°F or 122°F (25°C or 50°C), depending on the viscosity characteristics of the specific type and grade of asphalt emulsion.

For testing at 77°F (25°C), a sample is carefully stirred and conditioned for 30 minutes at the test temperature. The sample is then poured through a 0.85-millimeter sieve or strainer into a standard Saybolt Furol viscometer with a stopper placed in the orifice. The test is started by withdrawing the stopper and determining the time required for 60 milliliters of emulsion to flow through the orifice (see Figure 4.2). The time interval is called Saybolt Furol viscosity and is measured in seconds. The more viscous the material, the greater the length of time required for a given volume to flow through the orifice. Thus, a higher Saybolt Furol time is indicative of a higher sample viscosity.

For testing at 122°F (50°C), the sample is first heated to 122 ± 5°F (50 ± 3°C) and then poured through the strainer into the viscometer and brought to test temperature before stopper is removed and the flow is timed, as already described.

The Saybolt Furol viscosity test is suitable for asphalt emulsions that have a minimum flow time of 20 seconds.

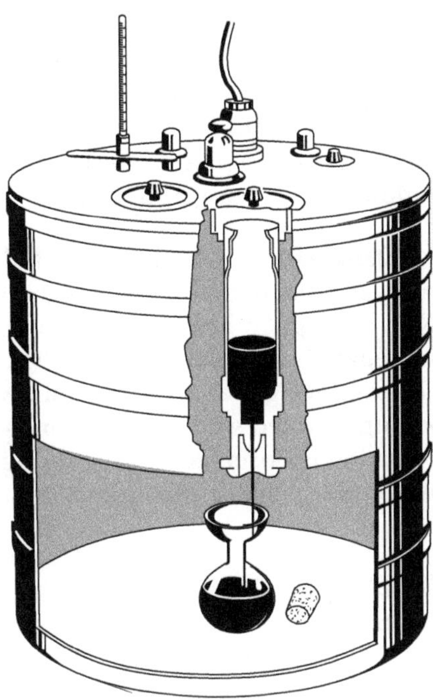

□ **Figure 4.2** Saybolt Furol Viscosity Apparatus

Demulsibility (ASTM D 244, D 6936)

The demulsibility test is used to identify rapid-setting asphalt emulsions by indicating the relative rate at which the colloidal asphalt particles will break when spread in thin films on soil or aggregate. Calcium chloride causes the minute asphalt particles present in these emulsified asphalts to coalesce. In the test, a solution of calcium chloride and water is thoroughly mixed with emulsified asphalt and then decanted over a sieve to determine how much the asphalt particles coalesce.

In rapid-setting emulsion tests, a very weak solution of calcium chloride and water is used. Specifications prescribe the concentration of the solution and the minimum amount of asphalt to be retained on the sieve. A high degree of "demulsibility" indicates an RS emulsion, which is expected to break almost immediately upon contact with the aggregate on which it is applied. When cationic rapid-setting emulsions are tested, a solution of sodium dioctyl sulfosuccinate in water is used in place of a calcium chloride solution.

Cement Mixing (ASTM D 244, D 6935)

The cement mixing test is used instead of the demulsibility test for slow-setting grades of asphalt emulsions. This test is specified for both the anionic and cationic types to ensure that the products are substantially immune to a rapid coalescence of asphalt particles in contact with fine-graded soils or dusty aggregates.

The cement mixing test is performed by stirring 100 milliliters of emulsion—which has been diluted with water to 55 percent residue—with 50 grams of high-early-strength Portland cement (Type III). After these ingredients are stirred for one minute, 150 milliliters of water is added and the mixture is stirred for an additional three minutes. The mixture is then washed over a No. 14 (1.40 millimeters) sieve and the percent of material retained on the sieve is determined. A low amount of retained weight (i.e., particles that have broken) is desirable for slow-setting grades.

Classification of Cationic Rapid-Setting Emulsions (ASTM D 244)

This test involves the coating of silica sand. The sand is first washed with hydrochloric acid and isopropyl alcohol. The emulsion is mixed with the sand for 2 minutes. At the end of the mixing period, if the uncoated portion is in excess of 50 percent of the total mix, it is considered a positive identification for a cationic rapid-setting emulsion.

Classification of Cationic Slow-Setting Emulsions (ASTM D 244)

This test is used if the result of the particle charge test is inconclusive. A weighed amount of washed and dried silica sand is mixed with a weighed amount of cationic slow-setting asphalt emulsion and mixed until the aggregate is completely coated. The amount of emulsion in the mix should be 5 percent by total weight of sand. The mix is cured for 24 hours and then placed in a beaker of boiling distilled water. After 10 minutes, the sample is placed on a level surface and the coating is observed. If the coating is in excess of 50 percent of the total mix, it is considered a positive identification of a cationic slow-setting emulsion.

Settlement and Storage Stability (ASTM D 244, D 6930)

The purpose of the settlement and storage stability tests is to detect the tendency of asphalt particles to "settle out" during storage of asphalt emulsions. The user is provided with an element of protection against separation of asphalt and water in unstable asphalt emulsions that may be stored for a period of time.

Either test is conducted by placing a 500-milliliter sample in a graduated cylinder, which is then stoppered and allowed to stand undisturbed. For the storage stability test, the sample must remain undisturbed at $77 \pm 5°F$ ($25 \pm 3°C$) for 24 hours. For the settlement test, the sample must remain undisturbed at $77 \pm 5°F$ ($25 \pm 3°C$) for 5 days. At the end of the storage period, small samples are then taken from the top and bottom parts of the cylinder.

Each sample is placed in an individual beaker and weighed. The samples are then heated to $325°F$ ($163°C$) for a total of 3 hours, allowed to cool to room temperature, and weighed. The weights obtained provide the basis for determining the difference, if any, between asphalt cement content in the upper and lower portions of the graduated cylinder, thus providing a measure of the settling of the asphalt particles from the emulsion.

Oversized Particles (Sieve) (ASTM D 244, D 6933)

This test is used to determine quantitatively the percent of asphalt cement present in the form of oversized particles. Such particles of asphalt might clog equipment and tend to provide non-uniform coatings of asphalt on aggregate particles.

This test is conducted by bringing the emulsion to the proper test temperature depending on its Saybolt Furol viscosity. Asphalt emulsions having a high viscosity (greater than 100 seconds at $77°F$ ($25°C$) or any emulsion required to be tested at $122°F$ ($50°C$) are tested at $122°F$ ($50°C$). The sample is poured through a No. 20 (850 micron) sieve. For anionic asphalt emulsions, the sieve and retained asphalt are rinsed with a mild sodium oleate solution. For cationic asphalt emulsions, distilled water is the rinse agent. After rinsing, the sieve and retained asphalt are dried in an oven and the relative amount of asphalt retained on the sieve is determined.

Coating Ability and Water Resistance (ASTM D 244)

This test has three purposes, to determine:

- the ability of an asphalt emulsion to coat the reference aggregate;
- the ability of the emulsion to withstand mixing; and
- the water resistance of the emulsion-coated aggregate (stripping).

The test is primarily used to identify medium-setting asphalt emulsions suitable for mixing with coarse-graded calcareous aggregates. This test is not applicable to rapid-setting or slow-setting asphalt emulsions.

The reference aggregate is coated with calcium carbonate dust and then mixed with the asphalt emulsion. About one-half of the mixture is then placed on absorbent paper for a visual inspection of the surface area of aggregate coated by the asphalt emulsion. The remainder of the mixture is sprayed with water and rinsed until the rinse water runs clear. This material is then placed on absorbent paper and inspected for coating. The test is then repeated and for this second run, the aggregate is coated with water before the emulsion is added, mixed, and then visually inspected for good, fair, or poor coating ability.

Density (ASTM D 244, D 6937)

The density (kilogram/liter or pound/gallon) is calculated by determining the mass of an asphalt emulsion in a standard measure of known volume. Results are reported to the nearest 0.01 pound/gallon at 77°F (0.01 kilogram/liter at 25°C).

Residue and Oil Distillate by Distillation (ASTM D 244, D 6997)

Distillation is used to separate the water from the asphalt. If the material contains oil, it will be separated along with the water. The relative proportions of asphalt cement, water, and oil in the emulsion can be measured after the distillation has finished. Additional tests may be run on the asphalt cement residue to determine its physical properties.

The ASTM D 6997 distillation test procedure for asphalt emulsions uses an aluminum alloy still and ring burners (Figure 4.3). Normally, the distillation is run at a temperature of 500°F (260°C) for 15 minutes under vacuum. Since the emulsion seldom approaches this temperature in the field, it should be noted that some residue properties could be altered, such as elastic properties given by polymer modification. Some agencies have changed the temperature and time at which the distillation test is run for these special products.

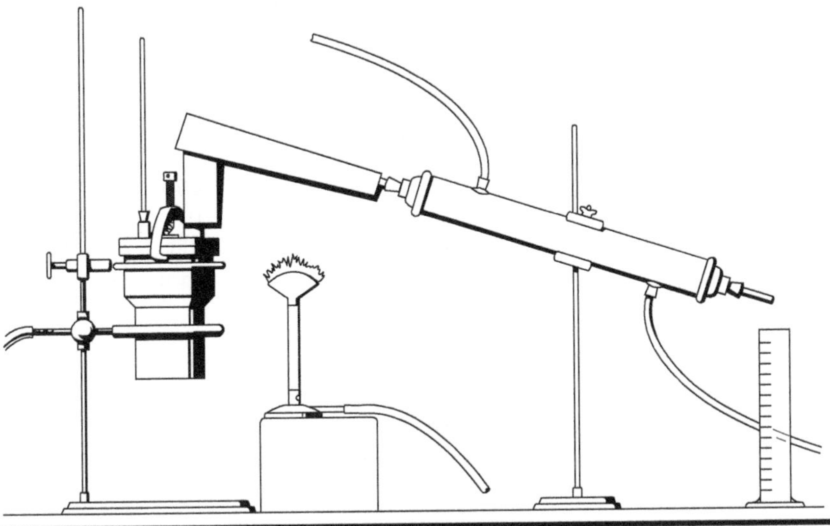

□ **Figure 4.3**   Distillation Test for Asphalt Emulsions

Residue by Evaporation (ASTM D 244, D 6934)

The evaporation procedure is carried out in an oven at a temperature of 325°F (163°C) for 3 hours. This evaporation procedure is sometimes used in lieu of a distillation procedure when difficulty arises in obtaining a representative residue sample. The evaporation procedure should not be used if a float test is to be run on the residue.

Breaking Index (no ASTM or AASHTO standards)

The breaking index test is designed to measure the speed of break of rapid-setting asphalt emulsions. Specified silica sand is added to 100 grams of the emulsion while stirring under controlled conditions of rate and temperature. The weight of the sand in grams needed to cause the emulsion to break (turn black and sticky) is considered the breaking index.

Residue Tests

The properties of the residual asphalt relate to field performance of the treatment in which the emulsion is used. The tests listed in this section are the most common tests performed on the residue. Sample preparation for these tests should be accomplished using residue directly from the distillation and evaporation procedures without reheating.

Penetration (ASTM D 5, AASHTO T 49)

The penetration test is one of the oldest empirical tests for measuring the consistency of asphalt. To conduct the penetration test, the asphalt residue is poured into a test container—usually a 3-ounce (88-milliliter) tin.

After a specified conditioning period, the sample is brought to the standard test temperature of 77°F (25°C) in a temperature-controlled water bath. The sample container is then placed in the penetrometer equipment.

A needle of prescribed dimensions is attached to the penetrometer and suspended directly over the asphalt binder sample. A 50-gram weight is attached to the needle's loading platform so that the total weight used for loading is 100 grams (50-gram weight plus needle weight of 50 grams). The penetrometer is lowered until the needle tip just contacts the surface of the asphalt binder. The load is then released, allowing the weighted needle to penetrate the asphalt binder for 5 seconds. The distance that the needle penetrates into the sample is reported as the penetration value. This distance is reported in units of 0.1 millimeter, or deci-millimeter (dmm). A typical penetration test consists of three measurements made on the sample that are averaged to provide a single test value. Between each measurement, the needle is replaced with a clean needle and the sample container is rotated so that a different portion of the sample is tested.

Ductility (ASTM D 113 and AASHTO T 51)

The ductility test procedure is performed by first assembling a standard V-shaped mold. The residue is then poured into the mold, slightly overfilling the mold. After a prescribed cooling period, the specimen is trimmed flush with the surface of the mold. The test specimen is then placed in the ductility water bath and conditioned to the desired test temperature, usually 77°F (25°C).

The specimen is then loaded into the ductility machine and one end of the specimen is pulled away from the other at a specified rate of speed, normally 5 centimeters per minute, until the thread of asphalt connecting the two parts of sample breaks. The elongation (expressed in centimeters) at which the thread of material breaks is designated as the ductility of the asphalt.

Float Test (ASTM D 139 and AASHTO T 50)

This test is commonly conducted on residue produced from high-float (HF) emulsions. The float test uses a semispherical float made from aluminum or aluminum-alloy that has an 11 millimeter diameter opening in the bottom of the float. The residue sample is poured into a brass collar designed to be screwed in place in the float (see Figure 4.4). After the sample cools in the collar, it is placed in the float and allowed to float in a water bath operating at 140°F (60°C). The time is measured from the instant the float is placed in the bath until water breaks through the asphalt residue.

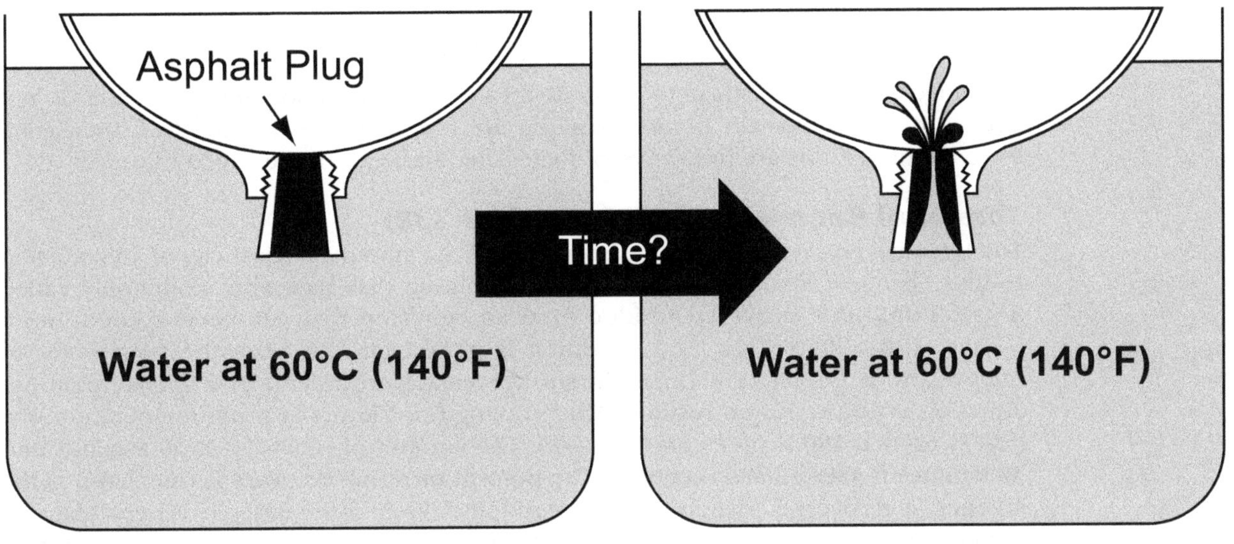

☐ **Figure 4.4**   Float Test

Elastic Recovery (ASTM D 6084 and AASHTO T 301)

The Elastic Recovery test is a procedure that uses the ductility test equipment. In the test, a specimen is prepared as in the ductility procedure, except that the side pieces are flat, rather than V-shaped. To conduct the test, a sample of asphalt residue is poured into the mold, slightly overfilling the mold. After a prescribed cooling period, the specimen is trimmed flush with the surface of the mold. The trimmed test specimen is placed in a water bath and conditioned to the desired test temperature, usually 77°F (25°C). After the conditioning period, the specimen is loaded into the ductility machine and one end of the specimen is pulled away from the other at a rate of 5 centimeters per minute until the specimen has stretched a prescribed length—either 10 or 20 centimeters.

Depending on the procedure, the specimen is then either immediately cut in the center or held in a stretched position for 5 minutes before cutting in the center. After 1 hour, the specimen is pushed back together until the cut ends touch. This final measurement is then used to calculate the recovery of the specimen. Higher recovery values are indicative of more elastic asphalt binders.

Force Ductility (AASHTO T 300)

The Force Ductility test uses the ductility test equipment. A specimen is prepared as in the ductility procedure, except that the side pieces are flat, rather than V-shaped.

The specimen preparation and test procedure is essentially identical to the elastic recovery test. The main differences between the tests are the following:

- A load cell is used to record the force required to pull the asphalt binder specimen.
- The test is continued to 30 centimeters at which point the test ends.
- Test temperature is usually 41°F (4°C).

In the test, a stress-strain curve is constructed. Typically, the stress at 30 centimeters is divided by the peak stress to generate a force ductility ratio. Specimens that do not stretch to 30 centimeters before breaking are considered to fail the test. Increasing force ductility ratios are indicative of increasing elasticity in the asphalt binder.

Torsional Recovery (California Test 332)

The torsional recovery test is a measurement of the amount of elasticity of an emulsion residue. The test involves imbedding an aluminum disk assembly, commonly called a spider, into the recovered residue from an emulsion that has been poured into a 3-ounce (88-milliliter) tin. The specimen is placed into an oven for 10 minutes to remove any air bubbles and break the surface tension around the disk to ensure proper adhesion of the emulsion residue. After cooling for 2 hours at room temperature, the disk is rotated 180 degrees and released. The amount of recovery at 30 seconds and 30 minutes is marked and recorded. The percent torsional recovery is calculated as the amount of recovery between the 30-second and 30-minute marks as a percentage of the circumference of the container. The percent torsional recovery is usually reported as an average of three specimens.

Toughness and Tenacity

In the test, a metal hemispherical head is embedded in hot asphalt to a depth of approximately 11 millimeters. After cooling to 77°F (25°C), the head is attached to a tensile test machine and pulled from the asphalt binder at a rate of 51 centimeters per minute. The load is measured throughout the test and a load-deformation curve is plotted. Toughness and Tenacity values are determined based on the area under different portions of the load-deformation curve.

Ring-and-Ball Softening Point (ASTM D 36 and AASHTO T 53)

In this test, asphalt residue samples are confined in small brass rings and loaded with a steel ball in the center of each ring. After the samples are prepared, the assembly is suspended in a beaker of water, glycerin, or ethylene glycol at 1 inch (25 millimeters) above a metal plate. The liquid is then heated at a prescribed rate. As the asphalt softens, the balls and the asphalt gradually sink toward the plate. At the moment the asphalt touches the plate, the temperature of the liquid is determined, and this is designated as the Ring-and-Ball (R&B) softening point of the asphalt residue.

Solubility (ASTM D 2042, AASTO T 44)

The solubility test is a measure of the purity of the asphalt residue. The portion of the residue that is soluble in carbon disulfide represents the active cementing constituents. Inert components—such as salts, free carbon, or non-organic contaminants—are insoluble. Due to the hazardous nature of carbon disulfide, trichloroethylene is used in the solubility test. Determining the solubility of an asphalt binder involves dissolving approximately 2 grams of asphalt in 100 milliliters of solvent, and then filtering the solution through a filter paper placed in a porcelain (Gooch) crucible with holes in the bottom to allow fluid to pass. The amount of material retained on the filter is determined by weighing and is expressed as a percentage of the original sample weight.

Absolute Viscosity (ASTM D4957 and AASHTO T 202)

The absolute viscosity test at 140°F (60°C) employs a capillary tube viscometer. Two types are most commonly used: the Asphalt Institute vacuum viscometer and the Cannon-Manning vacuum viscometer.

The viscometer is mounted in a thermostatically controlled, constant temperature bath. Water is often used as the liquid medium in the bath because the test temperature is below its boiling point. The asphalt emulsion residue is then poured into the large side of the viscometer until its level reaches the filling line. The filled viscometer tube is then placed into an oven at a prescribed temperature for a short time. Finally, the viscometer is placed in the constant temperature bath for a prescribed time to allow the system to reach an equilibrium temperature of 140°F (60°C).

At 140°F (60°C), the asphalt emulsion residue is too viscous to flow readily through a capillary tube viscometer. Therefore, absolute viscosity test equipment uses a vacuum pump to apply a consistent partial vacuum to the small side of the viscometer to induce flow.

After the bath, viscometer, and asphalt emulsion residue have stabilized at 140°F (60°C), the prescribed vacuum is applied and the time in seconds required for the asphalt binder to flow between the timing marks (delineating a section of capillary tube also called a "bulb") is precisely measured. Multiplying the measured time to flow through this bulb by the calibration constant for the particular bulb of the viscometer gives the value for viscosity in poises, the standard unit for measurement of viscosity.

The Cannon-Manning vacuum viscometer is normally used for residues from asphalt emulsions that are not modified by polymers or are not high-float (HF) emulsions. Residues from polymer-modified emulsions and high-float emulsions are non-Newtonian in nature and will not give consistent values between the two bulbs of a Cannon-Manning vacuum viscometer. For residues from polymer-modified emulsions, or high-float emulsions, the Asphalt Institute vacuum viscometer should be used with the results graphed and the viscosity selected at a shear rate of one reciprocal second (1 sec^{-1}). A Newtonian fluid would be a horizontal, or flat line, of viscosity graphed at various shear rates, while a non-Newtonian fluid is a sloped line. The viscosity of a Newtonian fluid is not affected by shear rate.

There are several different sizes of Asphalt Institute and Cannon-Manning vacuum viscometers with different size capillary tubes. The appropriate size viscometer should be selected based on the type of asphalt emulsion residue.

CHAPTER 5

SELECTING THE RIGHT TYPE AND GRADE

Successful performance of asphalt emulsions requires selecting the proper type and grade for the intended use. Guidelines presented in this chapter should help select the specific grade and type of emulsion to be used.

The first consideration in picking the right type and grade of emulsion is how the emulsion will be used. Is it for a plant mix (central or mixed-in-place), a recycled mix, or a prime coat application? Is it for some type of surface application, such as a fog seal, slurry seal, micro surfacing, or chip seal? Is it for a maintenance mix? Once this decision is made, other project variables must then be considered. Some other factors that affect the selection are:

- Climatic conditions anticipated during construction. The choice of emulsion grade, the design of mix or treatment, and the selection of construction equipment should be dictated by the conditions at the time of construction.
- Aggregate type, gradation, and availability.
- Contractor or construction equipment availability.
- Geographical location. The hauling distance and, in some cases, water availability are important considerations.
- Traffic control. Can traffic be detoured or only controlled through the work area?
- Environmental considerations.
- Proper application for pavement preservation or pavement distresses.
- Traffic type and volume.

While general guidelines can be given for selecting emulsions, laboratory testing is strongly recommended. There is no substitute for a laboratory evaluation of the emulsion and the aggregate to be used. Different types and quantities of emulsion should be tried with the aggregate to find the best combination for the intended use. An experienced technician can determine the type and grade of emulsion to be used.

General Emulsion Uses

Each grade of asphalt emulsion is designed for specific uses. They are described in general terms here. Emulsion nomenclature is described in Chapter 2. Table 5.1 shows the general uses of standard asphalt emulsion types and grades.

☐ Table 5.1: General Uses of Asphalt Emulsion

Type of Constructions	ASTM D 977, AASHTO M 208									ASTM D 2397, AASHTO M 140						
	RS-1	RS-2	HFRS-2, HFRS-2h	MS-1, HFMS-1	MS-2, HFMS-2	MS-2h, HFMS-2h	HFMS-2s	SS-1	SS-1h	CRS-1	CRS-2, CRS-2h	CMS-2	CMS-2h	CSS-1	CSS-1h	CQS-1H
Asphalt-Aggregate Mixtures:																
Plant Mix (Warm)[A]					X	X						X	X			
Plant Mix (Cold)																
Open-Graded Aggregate					X	X						X	X			
Dense-Graded Aggregate						X	X	X	X					X	X	
Sand						X	X	X	X					X	X	
Mixed-in-Place																
Open-Graded Aggregate				X	X	X						X	X			
Well-Graded Aggregate						X	X	X	X					X	X	
Sand						X	X	X	X					X	X	
Sandy Soil						X	X	X	X					X	X	
Asphalt-Aggregate Applications:																
Single & Multiple Chip Seals	X	X	X							X						
Sand Seal	X	X	X	X						X	X					
Slurry Seal									X						X	X
Micro Surfacing																X[E]
Sandwich Seal		X	X							X						
Cape Seal[F]		X	X						X	X					X	X
Asphalt Applications:																
Fog Seal	X			X[B]				X[C]	X[C]	X					X[C]	X[C]
Prime Coat					X[D]			X[D]	X[D]				X		X[D]	X[D]
Tack Coat/Bond Coat	X			X[B]				X[C]	X[C]	X					X[C]	X[C]
Dust Palliative	X[B]							X[C]	X[C]	X[B]				X[C]	X[C]	X[C]
Crack Filler								X	X					X	X	
Maintenance Mix:																
Immediate Use					X	X	X					X	X			
Stockpile					X		X					X				

[A]Other grades may be used where experience has shown that they give satisfactory performance.
[B]Diluted with water by manufacturer.
[C]Diluted with water.
[D]Mixed-in prime only.
[E]Polymer must be added during or prior to emulsification.
[F]Combination treatment, see Table 6.1.

Rapid-Setting Emulsions

The rapid-setting grades are designed to react quickly with aggregate and revert from the emulsion to the asphalt. They are used primarily for spray applications, such as chip seals, and sand seals. The RS-2, HFRS-2, and CRS-2 grades have higher viscosity to prevent runoff. Polymer-modified versions of these emulsions are routinely used where rapid adhesion is necessary. Examples include high traffic areas, where there is minimal traffic control, or where there is heavy truck traffic.

Medium-Setting Emulsions

Medium-setting grades are designed for mixing with graded aggregate. Because these grades are formulated not to break immediately upon contact with aggregate, they can coat a wide variety of graded aggregates. Mixes using medium-setting emulsions can remain workable from a few minutes to several months depending upon the formulation. Mixes are produced in pugmills and travel plants or can be road mixed. Medium-setting emulsions have been used in cold recycling applications.

Examples of medium-setting emulsions are MS-2, CMS-2, and HFMS-2. Nomenclature for medium-setting emulsions varies from state to state. Consultation with your local emulsion manufacturer is suggested for recommendations.

Polymer-modified versions of medium-setting emulsions may be used where additional stability or improved durability is needed or where improved water resistance is important.

Slow-Setting Emulsions

The slow-setting grades are designed for mixing stability. They are used with high fines content, dense-graded aggregates. The slow-setting grades have extended workability times to ensure good mixing with dense-graded aggregates. These mixes are not designed for stockpile storage. The slow-setting emulsions in mixing applications are generally used for dense-graded aggregate-emulsion bases, soil-asphalt stabilization, and for some recycling and slurry sealing.

All slow-setting grades have low viscosity that can be further reduced by adding water. When diluted, these grades can also be used for tack coats/bond coats, fog seals, and as dust palliatives.

Polymer-modified slow-setting emulsions may be used where additional mixture stability is needed or a better bond is necessary, the latter in the case of a tack coat/bond coat or fog seal.

Quick-Setting and Micro Surfacing Emulsions

The quick-setting grades are designed specifically for micro-surfacing and slurry seal applications when a quick curing time is necessary. This allows a quicker opening to traffic than the slow-setting slurry seal emulsions.

While a slurry seal is placed at the thickness of the largest aggregate in the gradation, micro-surfacing emulsions are polymer-modified and allow mixes to be placed at a greater thickness. The newly micro surfaced pavement can normally be opened to traffic in less than one hour after placement.

Laboratory evaluation is important to determine compatibility with job aggregates.

Guidelines for Successful Performance

The success with any type and grade of asphalt emulsion system is best ensured by strictly adhering to these steps:

- Conduct complete laboratory testing using the actual aggregate and emulsion that are to be used on the project.
- Select grades in conformance with Table 5.1.
- Adhere to the specifications and guidelines in this manual.
- Carefully handle the emulsion to prevent contamination, settlement of the asphalt particles, or premature coalescence.
- Contact the emulsion manufacturer's representative when special or unusual problems occur.

CHAPTER 6

TYPES OF SURFACE TREATMENTS

Asphalt Surface Treatments

Asphalt surface treatments are versatile, widely used forms of asphalt construction and maintenance. As the following sections indicate, there are several forms of surface treatments. When properly designed, they are economical, easy to place, and long-lasting. They seal and add life to road surfaces, but each type has one or more special purposes. A surface treatment is not a pavement. Rather, it provides a protective, waterproof cover over the underlying structure, which helps to resist traffic abrasion. It adds little or no load-carrying strength and, therefore, is not usually taken into account in computing the load-carrying capacity of a pavement. While a surface treatment can provide an excellent surface if it is used for the correct purpose, it is not a solution for all pavement problems. A clear understanding of the advantages and limitations of asphalt surface treatments is essential for best results. It is vital that a careful study of traffic requirements, along with an evaluation of the condition of the existing pavement, be made before deciding whether a surface treatment should be used.

Asphalt surface treatment is a broad term embracing several types of asphalt applications with or without a cover of mineral aggregate. Surface treatments are typically less than 1-inch (25-millimeters) thick. Asphalt emulsions used in these surface treatments are discussed in Chapter 5.

Surface treatments provide a waterproof cover over the existing pavement surface and provide resistance to abrasion by traffic. Specific functions of surface treatments are as follows:

- To provide long-lasting, economical surfaces for granular-base roads having light and medium traffic volumes. When polymer-modified emulsions and high quality aggregates are used, surface treatments can be used for higher volume traffic applications.
- To prevent surface water from penetrating granular bases or old pavements that have become weathered or cracked.
- To plug voids, and to coat and bond loose aggregate particles in pavement surfaces.
- To renew a surface and restore skid resistance to traffic-worn pavements in which the surface aggregate has become polished.
- To restore weathered surfaces and give new life to oxidized pavement surfaces.
- To provide temporary cover in cases of delayed pavement construction or planned stage construction.
- To control dust on low-volume roads.
- To promote adherence of subsequent asphalt courses to granular bases (prime coats).

- To ensure a bond between subsequent pavement layers being placed or overlays of existing surfaces (tack/bond coats).
- To improve the aesthetics of an existing pavement.

The treatments discussed in this chapter are summarized in Table 6.1.

☐ **Table 6.1: Asphalt Emulsion Surface Treatments**

Treatment Type	Description and Uses	Standard AASHTO Grades	Construction Hints
Fog Seal	A light spray application of binder applied to the surface of a chip seal, an open-graded mix, or a weathered hot mix surface. Provides some crack sealing, reduces raveling, and enriches weathered surfaces.	RS-1, SS-1, SS-1h, CRS-1, CSS-1, CSS-1h	Spray-applied with or without sand cover. Dilute the emulsion with water to help achieve coverage without adding excess binder.
Sand Seal	Restores uniform cover. In city street work, improves street sweeping, traffic line visibility. Enriches dry, weathered pavement; reducing raveling.	CRS-1, CRS-2, RS-1, RS-2, MS-1, HFMS-1, HFRS-2	Spray-applied with sand cover. Roll with pneumatic roller. Avoid excess binder.
Chip Seal	Single most utilized maintenance method. Produces an all-weather surface, renews weathered pavements, improves skid resistance, lane demarcation, and seals pavement.	CRS-2, CRS-2h, RS-2, CRS-2P, CRS-2L, HFRS-2, HFRS-2h	Spray-applied. Many types of textures available. Multiple applications possible. Keys to success: coordinate construction; use hard, clean aggregate and properly calibrated spray equipment; control traffic.
Sandwich Seal	Improves skid resistance, seals pavement.	RS-2, CRS-2, HFRS-2 (usually polymer modified)	Spread large aggregate. Spray apply emulsion, and then cover with smaller aggregate to lock in larger aggregate. Clean aggregate required.
Slurry Seal	Used in airport and city street maintenance where loose aggregate cannot be tolerated. Seals, fills minor depressions, provides an easy-to-sweep surface. The liquid slurry is machine-applied with a sled-type box containing a rubber-edged strike-off blade.	CQS-1h, CSS-1h, SS-1h	Test the aggregate and emulsion mix to achieve desired proportions, workability, setting rate, and durability. Calibrate equipment prior to starting the project.
Micro Surfacing	High performance resurfacing used in highway, city street, and airport maintenance where a durable, friction-resistant resurfacing is required. Rapid roadway surface correction. Special rut-filling application boxes and stringent design criteria permit filling wheel ruts up to 1.5 inches in depth in one pass.	CQS-1h (polymer-modified)	A mix design should be required. Calibrate equipment prior to starting the project. Experienced personnel required for proper application.
Cape Seal	Combines a single chip seal with a slurry seal. Provides the rough, knobby surface of a chip seal to reduce hydroplaning, yet has a tough matrix for durability. Test track data indicate better studded tire damage resistance than a chip seal. Friction values can be higher than conventional hot mix asphalt.	CQS-1h, CSS-1h, SS-1h, RS-2, CRS-2, CRS-2h, CRS-2P, CRS-2L, HFRS-2, HFRS-2h	Apply a single aggregate chip seal. After curing, broom loose material and apply the slurry seal. Have the strike-off ride on the rock surface to form the matrix. Avoid excess stone that can cover the desired knobby texture of the chips.

Site Evaluation and Preparation

Some pavements need minor repairs before they are resurfaced, while others may require complete rehabilitation such as recycling or reclamation. Before any work is started, a thorough surface examination is made to determine needed repairs and proper surface treatment. Corrective action needs to be taken to address base failures, drainage issues, variations in width, cross section, and profile. Repair potholes, cracked areas, depressions, slick or bleeding asphalt areas, absorbent areas, and other surface defects before treatment is begun.

When surface repairs such as patching are performed, sufficient time should be allowed before treating the surface to ensure consolidation under traffic. The asphalt may not adhere to the pavement, unless the surface to be covered is completely clean. A vacuum broom (Figure 6.1) is recommended to pick up both dust and loose particles, especially in urban areas, but if one is not available, a rotary power broom can be used (Figure 6.2). When brooms are used, flushing with water may be necessary to meet clean air standards or to remove caked-on materials. Power brooms are also used to remove loose particles after the treatment is completed and the asphalt has properly cured. Only light broom pressure should be used so that the aggregate particles embedded in the asphalt membrane are not dislodged.

☐ **Figure 6.1**   Vacuum Broom

☐ **Figure 6.2**   Rotary Power Broom

Spray-Applied Seals

Fog Seal

A fog seal is a light application to an existing surface of a slow setting asphalt emulsion diluted with water. It can be diluted in varying proportions—up to 1 part emulsion to 5 parts water, but in most cases, a one-to-one dilution is used. Grades of asphalt emulsion normally used for this purpose are SS-1, RS-1, SS-1h, CSS-1, CRS-1, or CSS-1h.

A fog seal can be a valuable maintenance aid when used for its intended purpose. It is not a substitute for an asphalt-aggregate surface treatment. It is used to renew old HMA pavement surfaces that have become dry and brittle with age, to seal small cracks and surface voids, and to inhibit raveling.

The fairly low-viscosity diluted emulsion flows easily into the cracks and surface voids. It also coats aggregate particles on the surface.

This corrective action prolongs pavement life and may delay the need for major maintenance or rehabilitation.

The total quantity of fog seal used is approximately 0.10–0.15 gallon/square yards (0.45–0.70 liter/square meters) of diluted material. Exact quantities are determined by the surface texture, dryness, and degree of cracking or raveling of the pavement on which the fog seal is to be sprayed. Over application must be avoided because it will result in asphalt pickup by vehicles and possibly create a slippery surface. If excessive emulsion is applied, a light dusting of the affected area with fine sand may remedy the problem.

Traffic must be kept off the fog seal until the emulsion breaks and is substantially absorbed into the existing surface. This curing period may range from one hour in hot, dry conditions to as much as three hours or longer in cool, humid conditions.

Rejuvenating Emulsion

Asphalt consists of two fractions: asphaltenes; the solid fraction, and maltenes, the liquid fraction. During the aging process, the maltene oxidizes and the ratio of maltenes to asphaltenes is reduced. The aging process causes asphalt mixes to become dry and brittle. Asphalt rejuvenating emulsions can be used to improve the properties of the aged asphalt.

Rejuvenating emulsions can be applied as a fog seal when the level of cracking is greater than what would normally be used as a criterion for typical chip or fog seals.

Rejuvenating emulsions are also used as recycling agents to restore the aged binder in recycled mixes. See Chapter 9.

Dust Palliative

A dust palliative is the application of a diluted asphalt emulsion directly on an unpaved surface as a dust control agent. SS-1, SS-1h, RS-1, CRS-1, CQS-1h, CSS-1, or CSS-1h diluted with 5 or more parts of water by volume is used. The diluted material is sprayed with an asphalt distributor in repeated small applications as required. The application rates are approximately 0.1–0.5 gallon/square yard (0.45–2.3 liters/square meter). The total amount applied depends on the condition of the existing surface. Some penetration of the emulsion is expected, and with a more open surface, more emulsion can be applied. As with other spray treatments without cover material, small test sections are recommended to determine the best application rate for the existing conditions.

Chip Seal

A chip seal (sometimes referred to as a bituminous surface treatment) is used to protect the pavement from the deteriorating effects of sun and water. A secondary benefit of a chip seal is an increase in the skid resistance of the pavement surface. This occurs because the cover aggregate increases the surface texture of the pavement. A single bituminous surface treatment is used as a wearing and waterproofing course. It consists of a sprayed application of asphalt immediately covered by a single layer of uniform-size aggregate. The thickness of the treatment approximates the nominal maximum-size aggregate particles used in the particular surface treatment.

When a spreader drops a one-size cover aggregate on an asphalt film, the particles will lay in an unarranged position. After compaction, the particles will become oriented into their densest position with about 20 percent voids between the particles. It is desirable to fill these voids two-thirds to three-fourths full with asphalt emulsion. A typical design will call for 70 percent of the voids to be filled. As the water evaporates from the asphalt emulsion, the residual asphalt should, under average conditions, fill 55–60 percent of the voids between the aggregates.

A multiple treatment provides a wearing and waterproofing course of greater durability than a single treatment. It consists of two or more alternate applications of asphalt and aggregate. For each succeeding course, the nominal top size of cover aggregate should be not more than one-half the size of that for the previously placed course. No allowance is made for wastage. Also, after the first course, no correction is made for underlying surface texture. The total asphalt quantity required for all courses should be considered as a whole. For two course application 40 percent of the total should be applied for the first application and the remaining 60 percent applied for the second application. In a triple treatment, 30 percent of the total should be applied for the first application, 40 percent for the second application, and 30 percent for the third application. In multiple treatments, the first course of cover aggregate generally determines the thickness while subsequent courses partially fill the upper voids in the previously placed courses.

Materials

▶ Asphalt Emulsion

The selection of the proper type and grade of emulsion for a chip seal must include the following considerations:

- Temperature of the surface to which the treatment will be applied
- Air temperature
- Type and amount of traffic
- Condition of the surface
- Type and condition of the aggregate to be applied

The correct amount, type, and grade of emulsion for the chip seal will:

- be fluid enough to spray and cover the surface uniformly, yet viscous enough to remain in a uniform layer and not puddle in depressions or run off the crown at grade;
- after application, retain the required consistency to wet the applied aggregate;
- cure and develop adhesion quickly;
- after rolling and curing, hold the aggregate tightly to the road surface to prevent dislodging by traffic; and
- not bleed or strip.

The grades of emulsions most suitable for chip seals are listed in Table 6.1.

▶ Aggregate

Most aggregates, such as gravel, crushed stone, and crushed slag, can be successfully used for cover. The aggregate selected, however, must meet certain requirements of size, shape, cleanliness, and surface properties.

Aggregate should be as uniform as (a single size) economically practicable so that the surface treatment will have essentially only one layer of aggregate. Generally, the largest size should be no more than twice the smallest size, with a reasonable tolerance in dimensions to allow for economical production. Aggregates having a maximum size of less than 1/2 inch (13 millimeters) in diameter provide a quieter and smoother riding surface than larger ones.

The ideal aggregate shape is cubical. A large number of flat or elongated particles is undesirable because they may be completely covered by asphalt applied in a sufficient quantity to hold the cubical particles.

Good adhesion between aggregate and the residual asphalt is essential to a chip seal's success. It is critical that the aggregate be clean. If the particles are coated with dust, silt, or clay, the coating forms a film that prevents asphalt-aggregate adhesion. Slightly damp, versus completely dry, aggregate adheres better when emulsions are used.

▶ Chip Seal Design

After the appropriate type and grade of emulsion is selected for proper compatibility with the aggregate, there are several theoretical procedures for determining the quantity of emulsion and cover aggregate. These usually involve determining the average least dimension, the bulk specific gravity, and voids (AASHTO T 19) of the cover aggregate. Mathematical calculations, coupled with laboratory testing, are typically used to determine the required quantities of emulsion and aggregate. Design procedures are available from ASTM and AASHTO; however, many state DOTs have their own procedures. The goal is to have aggregate particles approximately 70 percent embedded in emulsion. Adjustments must be made to account for the traffic volume, the absorption of the emulsion in the existing pavement and the cover aggregate, the texture of the existing pavement, and the characteristics of the cover aggregate (size, shape, and gradation). See Tables 6.2, 6.3, 6.4, and 6.5.

☐ **Table 6.2: Quantities of Emulsion and Aggregate for Single Surface Treatment** [1, 2, 3, 4, 5]

Nominal Aggregate Size	Size No.	Aggregate Quantity lb/yd² (kg/m²)	Emulsion Quantity gal/yd² (l/m²)	Emulsion Grade
3/4 to 3/8 in. (19.0 to 9.5 mm)	6	40–50 (22–27)	0.40–0.50 (1.8–2.3)	RS-2, CRS-2, CRS-2P, CRS-2L, HFRS-2
1/2 in. to No. 4 (12.5 to 4.75 mm)	7	25–30 (14–16)	0.30–0.45 (1.4–2.0)	RS-2, CRS-2, CRS-2P, CRS-2L, HFRS-2
3/8 in. to No. 8 (9.5 to 2.38 mm)	8	20–25 (11–14)	0.20–0.35 (0.9–1.6)	RS-2, CRS-2, CRS-2P, CRS-2L, HFRS-2
No. 4 to No. 16 (4.75 to 1.18 mm)	9	15–20 (8–11)	0.15–0.20 (0.7–0.9)	RS-1, MS-1, CRS-1, HFRS-2
Sand	AASHTO M-6	10–15 (5–8)	0.10–0.15 (0.5–0.7)	RS-1, MS-1, CRS-1, HFRS-2

[1] These quantities of emulsion cover the average range of conditions that include primed granular bases and old pavement surfaces. The quantities and types of materials may be varied according to local condition and experience.

[2] The lower application rates of emulsion listed in the above table should be used for aggregate gradations on the fine side of the specified limits. The higher application rates should be used for aggregate gradations on the coarse side of the specified limits.

[3] It is important to adjust the emulsion quantity for the surface condition of the road, increasing it if the road is absorbent, badly cracked, or raveled, and decreasing it if the road is flushed with asphalt (see Table 6.3).

[4] It is important to adjust the emulsion quantity for traffic count and conditions. An increase in traffic will mean a decrease in emulsion content.

[5] These aggregate quantities are based on a specific gravity of 2.65. If the aggregate specific gravity is outside of the range 2.55–2.75, adjustments should be made by the engineer.

☐ **Table 6.3: Correction for Surface Condition**

Pavement Texture	Correction**	
	gal/yd^2	L/m^2
Black, flushed asphalt	−0.01 to −0.06	−0.04 to −0.27
Smooth, non-porous	0.00	0.00
Absorbent—slightly porous, oxidized	0.03	0.14
—slightly raveled, porous, oxidized	0.08	0.27
—badly raveled, porous, oxidized	0.09	0.40

This correction must be made from observations at the job site.

☐ **Table 6.4: Quantities of Asphalt and Aggregate for Double Surface Treatment**

	Nominal Aggregate Size	Size No.	Aggregate Quantity lb/yd^2 (kg/m^2)	Emulsion Quantity gal/yd^2 (L/m^2)
1/2 in. (12.5 mm) Thick 1st Application*	3/8 in. to No. 8 (9.5 to 2.38 mm)	8	25–35 (14–19)	0.20–0.30 (0.9–1.4)
2nd Application	No. 4 to No. 16 (4.75 to 1.18 mm)	9	10–15 (5–8)	0.30–0.40 (1.4–1.8)
5/8 in. (15.9 mm) Thick 1st Application*	1/2 in. to No. 4 (12.5 to 4.75 mm)	7	30–40 (16–22)	0.30–0.40 (1.4–1.8)
2nd Application	No. 4 to No. 16 (4.75 to 1.18 mm)	9	15–26 (8–11)	0.40–0.50 (1.8–2.3)
3/4 in. (19.0 mm) Thick 1st application*	3/4 to 3/8 in. (19.0 to 9.5 mm)	6	40–45 (22–27)	0.35–0.50 (1.6–2.3)
2nd Application	3/8 in. to No. 8 (9.5 to 2.36 mm)	8	20–25 (11–14)	0.50–0.60 (2.3–2.7)

*A penetrating prime may be used on untreated granular (stone) base prior to chip seal application.

☐ **Table 6.5: Quantities of Asphalt and Aggregate for Triple Surface Treatment**

	Nominal Aggregate Size	Size No.	Aggregate Quantity lb/yd^2 (kg/m^2)	Emulsion Quantity gal/yd^2 (L/m^2)
1/2 in. (12.5 mm) Thick 1st Application*	3/8 in. to No. 8 (9.5 to 2.38 mm)	8	25–35 (14–19)	0.20–0.30 (0.9–1.4)
2nd Application	No. 4 to No. 16 (4.75 to 1.18 mm)	9	10–15 (5–8)	0.25–0.35 (1.1–1.6)
3rd Application	No. 4 to No. 100 (4.75mm to 150mm)	10	10–15 (5–8)	0.20–0.30 (0.9–1.4)
5/8 in. (15.9 mm) Thick 1st Application*	1/2 in. to No. 4 (12.5 to 4.75 mm)	7	30–40 (16–22)	0.20–0.30 (0.9–1.4)
2nd Application	3/8 in. to No. 8 (9.5 to 2.36mm)	8	15–20 (8–11)	0.30–0.40 (1.4–1.8)
3rd Application	No. 4 to No. 16 (4.75 to 1.18mm)	9	10–15 (5–8)	0.20–0.30 (0.9–1.4)
3/4 in. (19.0 mm) Thick 1st application*	3/4 in. to 3/8 in. (19.0 to 9.5 mm)	6	35–45 (19–25)	0.25–0.35 (1.1–1.6)
2nd Application	3/8 in. to No. 8 (9.5 to 2.36 mm)	8	20–30 (11–14)	0.30–0.40 (1.4–1.8)
3rd Application	No. 4 to No. 16 (4.75 to 1.18 mm)	9	10–15 (5–8)	0.25–0.35 (1.1–1.8)

*A penetrating prime may be used on untreated granular (stone) base prior to chip seal application.

Supplemental Mix Design Tests

Vialit Test

The Vialit procedure has been found quite useful in the design of chip seals typically using a rapid-setting emulsion. The emulsion is poured at its design application rate onto a metal plate and allowed to flow to a uniform film thickness. Aggregate is applied to the binder-covered plate and embedded by rolling. After it is cured, treated with water, and dried again, the plate is turned upside down. A 500 gram steel ball is dropped from a specified height three times on the reverse side of the plate and the loss of the aggregate is measured. There are various ways of reporting the results. This test is particularly helpful using the actual job aggregate under the field humidity and temperature conditions.

Sweep Test (ASTM D 7000)

The sweep test measures the curing performance of the asphalt emulsion and aggregate by simulating the brooming of a chip seal in a laboratory. Asphalt emulsion is applied to an asphalt felt disk at a prescribed amount, followed by aggregate being applied and embedded into the asphalt emulsion. The sample is conditioned at a prescribed temperature and time period before testing. A nylon brush attached to a stand mixer abrades the sample for one minute. After one minute the test is stopped and any loose aggregate is removed and the percent loss is calculated.

Equipment

Distributor

The distributor (see Figure 6.3) consists of a truck- or trailer-mounted insulated tank ranging in capacity from 800 to 5500 gallons (3000–20,800 liters), which applies the emulsion uniformly and in specified quantities. Most distributors are equipped with a heating system that will maintain the emulsion at the proper spraying temperature. A system of spray bars with nozzles applies emulsion to the road surface. These spray bars cover widths up to 30 feet (10 meters) in one pass.

○ **Figure 6.3** Asphalt Distributor

Although the methods of maintaining pressure may vary, all distributors use gear-type pumps to deliver emulsion to the spray bar. On some distributors, the pressure is governed by a variable-speed pump. On other distributors, it is controlled with a constant speed pump and a pressure-relief valve. The correct pump speed or pressure is that which creates the proper spray fan out of the nozzle. Low pressure may result in streaking, while high pressure may distort the fan spray. The manufacturer's equipment manual should be referenced to determine the volume-per-minute discharge for each nozzle size. A tachometer is used as an aid in maintaining uniform spreader speed. Newer distributors have interlocks between the asphalt pump and the forward speed of the distributor. As the distributor changes speed, the pump output is adjusted automatically to maintain a constant application rate.

The circulating system:

- fills the distributor tank;
- circulates material through the bar and tank;
- pumps material through the bar or hand spray;
- draws material back to the tank from the bar or hand spray;
- pumps material from the tank to outside storage; and
- transfers material from one storage tank to another.

One of the most important parts of the distributor is the spray bar. To achieve good results, the correct size nozzles for the type and grade of emulsion and the application rate must be selected. Before use, nozzles should be checked for damage and proper setting.

The angle of the nozzle openings (see Figure 6.4) must be adjusted so that the spray fans do not interfere with each other. This angle varies according to the manufacturer but is typically between 15 and 30 degrees (0.26 and 0.52 radians). It is important that all nozzles be set at the proper angle within close tolerances.

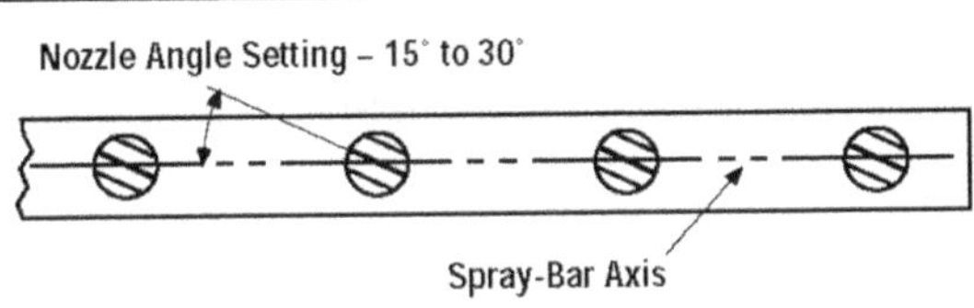

Note. On occasion, some operators will set end nozzles at a different angle (60 to 90 degrees with respect to the spray-bar) in an attempt to obtain a good edge. This practice should NOT be permitted as it will produce a fat streak on the edge and rob the adjacent spray fan of the lap from this nozzle. A curtain on the end of the bar or a special end-nozzle with all nozzles set at the same angle will provide more uniform coverage and make a better edge.

Figure 6.4 Proper Nozzle Angle Setting

The height of the spray bar above the pavement surface must be adjusted to ensure uniformity of the emulsion spread. It is important, too, that the correct height be maintained throughout the application. Incorrect spray bar height results in non-uniform coverage. If the spray bar is too high, wind distortion of the spray fans may occur.

For example, the best results with 4-inch (100-millimeter) nozzle spacing are achieved with a triple coverage of the spray fans (see Figure 6.5). But in order to get triple coverage with 6-inch (150-millimeter) nozzle spacing, the height of the spray bar would be so great that it could be subject to wind distortion. In such cases, a double-coverage pattern is used.

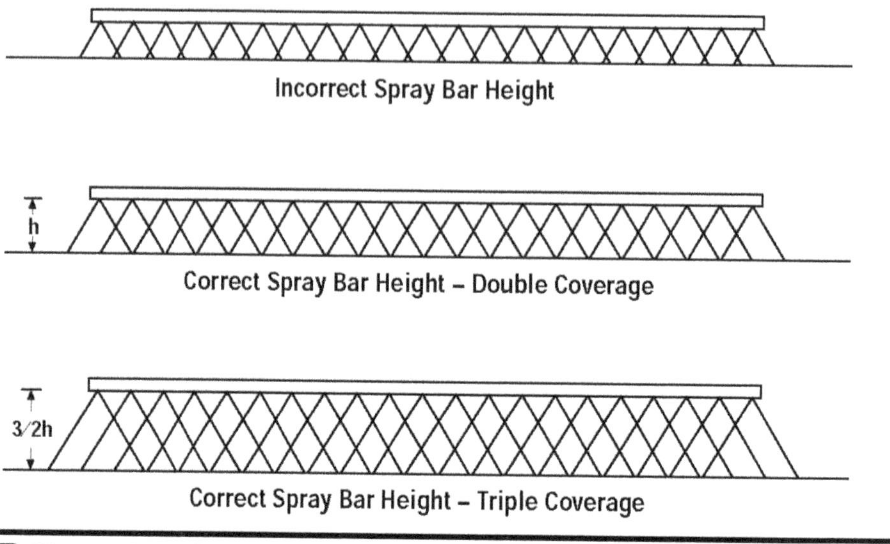

Figure 6.5 Spray Bar Height Must Be Set for Proper Coverage

Setting end nozzles at a different angle (60 to 90 degrees with respect to the spray bar) in an attempt to obtain a good edge should NOT be permitted. Rather, special half-spray end nozzles will provide more uniform coverage and make a better edge.

For best results, the height of the spray bar above the road surface should not vary more than 1/2 inch (13 millimeters) during application. Special steps must be taken to ensure this maximum permissible variance as the load lightens on the springs of the distributor truck. Many distributors have mechanical controls to maintain the proper height.

Distributor application rates can be verified by using this formula:

$$R = TM/WL$$

where:

R = Rate of application, gallons/square yard (liter/square meter)
T = Total gallons (liters) applied
W = Width of spread, yards (meters)
L = Length of spread, yards (meters)
M = Multiplier for correcting emulsion volume to basis of 60°F (15.6°C) (from Table B-1 in Appendix B).

Aggregate Spreader

The function of the aggregate spreader is to apply a uniform aggregate cover at a specified rate over the freshly sprayed emulsion. Aggregate spreaders for use in chip seal construction consist of three basic types:

- A tailgate spreader, which is connected directly to the tailgate of the truck.
- A mechanical spreader, which is also connected to the tailgate of the truck but has wheels to support the aggregate spreader.
- A mechanical, self-propelled spreader, which pulls the truck and spreads the aggregate.

There are several types of tailgate spreaders. The simplest is the vane spreader, as shown in Figure 6.6. It consists of a steel plate with a series of vanes, which spread the aggregate over the desired width on the road surface. Some tailgate spreaders consist of a hopper with a feed roller activated by small wheels that contact truck wheels (see Figure 6.7).

❑ **Figure 6.6** Tailgate Vane Spreader

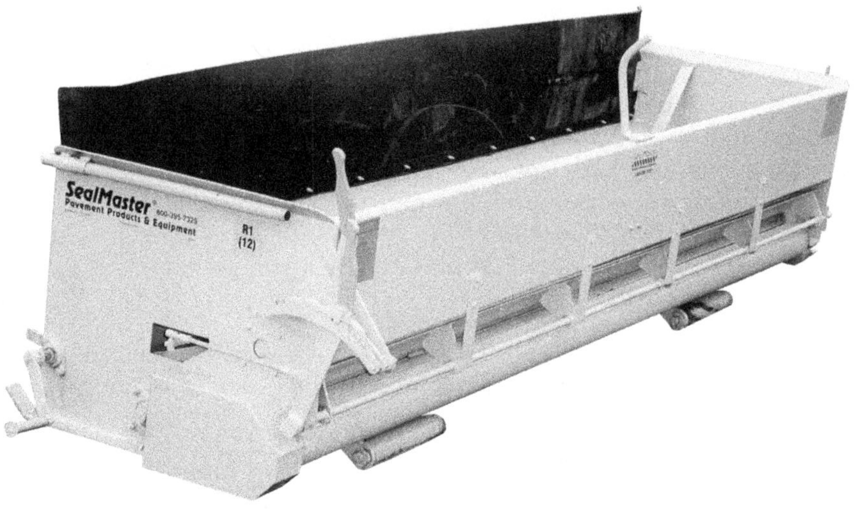

❑ **Figure 6.7** Truck-Attached Mechanical Spreader

Truck-mounted mechanical spreaders are hoppers on wheels, attached to and propelled by backing aggregate trucks. Hoppers usually contain augers to distribute the aggregate the full width of the box. The wheels support the spreader and operate the auger and a feed roller in the bottom of the hopper. The feed roller ensures a uniform spread of material onto the road surface. There is an adjustable hitch on the front of the spreader, which attaches to the aggregate truck. Mechanical spreaders also have controls to regulate the feed gates, the feed roll, the auger, and the truck hitch height.

A self-propelled spreader as shown in Figure 6.8 provides a uniform and continuous distribution of the aggregate, and makes high production rates possible. The self-propelled spreader attaches to the truck as it backs into the rear of the machine. The spreader then pulls the truck. A receiving hopper is located at the rear of the spreader. The aggregate is moved from the receiving hopper to the spreading hopper by conveyor belts. By adjusting the speed of each conveyor belt, the flow of aggregate into the spreading hopper can be set to match the flow from the receiving hopper.

○ **Figure 6.8** Self-Propelled Mechanical Spreader

With individually controlled aggregate flow gates, the operator can utilize as little as 6 inches (150 millimeters) of the width of the spreader to as much as the complete width of the spreader. This facilitates constructing radii, intersections, and variable roadbed widths. The aggregate, when dropping into the spreading hopper, usually passes through a scalping screen. This allows the removal of any oversize material. Aggregate drops into the spreading hopper and flows over a spread roller and is then deposited onto the emulsion-coated surface.

The calibration and adjustment of all types of aggregate spreaders should be made according to manufacturers' instructions and operation manuals. Some additional suggestions to ensure good results:

- The engine tachometer can be used to maintain uniform spreader-speed.
- Monitor aggregate spread rates by marking the length that each truckload covers.
- Calibration of the spreader is made by laying a section of cloth, building paper, or a shallow box that measures 1 square yard (1 square meter) on the pavement by passing over it with the spreader. The cloth, paper or box is then carefully lifted, and the aggregate is weighed.

Rollers

The seating of aggregate particles is an important part of the chip seal operation. There are several types of rollers, but pneumatic tire rollers are normally used.

A self-propelled pneumatic-rubber tire roller (Figure 6.9) with tire pressures in the range of 60–90 pounds per square inch (415–620 kiloPascals) is recommended for use on asphalt-aggregate surface treatments. The rubber tires reorient and seat the aggregate particles without crushing. The use of steel-wheel rollers is discouraged because they bridge uneven surfaces and may crush the aggregate.

○ **Figure 6.9** Pneumatic-Tire Roller

Cleaning Equipment

Unless the surface to be covered is completely clean, the asphalt may not adhere to the pavement. It is therefore necessary to clean the surface before spraying the asphalt emulsion. A vacuum broom (Figure 6.1) is recommended for picking up both dust and loose particles; if one is not available, a rotary power broom (Figure 6.2) can be used. When brooms are used, flushing with water may be necessary to meet clean air standards or to remove caked-on materials.

Power sweepers or brooms are also used to remove loose particles after the treatment is completed and the asphalt has properly cured. Only light broom pressure should be used so that aggregate particles embedded in the asphalt membrane are not dislodged. On a newly constructed chip seal, it is advisable to broom during cooler periods of the day to prevent chip rollover.

Trucks

Enough trucks must be available to ensure that the chip seal operation proceeds without interruption. Frequent stops and starts may cause variations in emulsion spray distribution, rate of aggregate cover, or both, and result in a non-uniform surface. By staggering backing patterns, trucks can also be used to roll the finished chip seal to help embed the aggregate before the regular rolling is begun. Truck speed and turning should be carefully controlled to prevent dislodging chips.

Construction

Sequence of Operations

1. Patch potholes and repair damaged areas in the existing pavement. Allow patch to cure. If a coarse patch mix is used, fog sealing may be advisable.
2. Clean the surface with a vacuum sweeper or rotary broom or by another approved method.
3. Spray the asphalt emulsion at the specified rate and proper temperature.
4. Spread the cover aggregate at the specified rate immediately behind the emulsion spray application (emulsion still brown in color) to achieve maximum possible chip wetting.
5. Roll the cover aggregate to orient and seat particles.

Figure 6.10 shows a proper chip seal operation. For a double or triple treatment, steps 3 through 5 will be repeated.

☐ **Figure 6.10** Chip Seal Operation

All equipment must be in proper working order before construction begins. An adequate supply of aggregate should be available on the job site, or scheduled for delivery at proper intervals with adequate haul trucks to permit continuous spreading operations. The required quantity of asphalt emulsion should also be stored or scheduled to arrive promptly at the job site to prevent delays in construction. A traffic control plan should be developed.

Transverse Joints

Rough and unsightly transverse joints can be avoided by starting and stopping the emulsion and aggregate spread on building paper. The paper is placed across the lane to be treated so that the forward edge is at the desired joint location. The distributor—traveling at the correct speed for the desired application rate—starts spraying on the paper so that the spray bar makes a full, uniform application when reaching the exposed surface. A second length of building paper is placed across the lane at the predetermined cutoff point for the distributor. This procedure gives a straight, sharp transverse joint. After the aggregate spreader has passed over it, the paper should be removed and disposed of properly.

For the next application, the leading edge of the paper is placed within a 1/2 inch (13 millimeters) of the cutoff line of the previously laid treatment. This prevents a gap between the two spreads.

Longitudinal Joints
To prevent aggregate from building up on the longitudinal joint, the edge of the aggregate spread should coincide with the edge of applied emulsion. The longitudinal joint should be along the centerline of the pavement being treated. An established line helps to ensure a straight longitudinal joint.

Precautions
Most problems with chip seals are caused by failure to adhere to common-sense construction practices. Even when the highest quality materials are used, inferior performance may result unless strict guidelines are followed. Shortcuts will likely result in poor performance and increased maintenance.

Chip seal operations should not be carried out during periods of cold or wet weather. Conventional guidelines recommend air temperatures of at least 50°F (10°C) in the shade and rising. Some specifications require the temperature of the road surface be above 70°F (20°C) before an emulsion can be applied. The emulsion may not break or cure properly at lower temperatures and may not satisfactorily retain the cover aggregate. Chip seals should not be constructed in the rain, when rain is threatening, or on wet pavement. The water may cause a loss of the partly cured emulsion from the cover aggregate.

Specially formulated emulsions have been developed, which can be used at cooler temperatures or at night. A local asphalt emulsion supplier should be contacted for specific recommendations.

Best Practices
The following simple safeguards will greatly increase the chance of success when a chip seal is constructed:

- Assure the existing pavement structure is strong enough to support expected traffic.
- Perform the design with materials representative of those to be used on the job.
- Use aggregate which is free from dust and slightly damp for the best results.
- Inspect and calibrate the equipment to ensure proper operation.
- Select the proper type and weight of rollers.
- Follow proper construction techniques.
- Provide proper traffic control.
- Perform the work in weather conditions suitable for the type and grade of emulsion selected.

Adhering to the simple safeguards will help prevent problems. Three of the most common problems and their causes are:

Streaking

Streaking is the non-uniform application of the asphalt emulsion on the road surface (Figure 6.11). Longitudinal streaking shows up as alternating lean and heavy bands of asphalt running parallel to the centerline of the pavement. Streaking not only leaves an unsightly appearance, it can also greatly reduce service life through the loss of cover aggregate. A single centerline streak may be caused by too little or too much overlap at the joint between two applications.

☐ **Figure 6.11** Longitudinal Streaking

Causes of longitudinal streaking include:

- Improper spray bar height causing incorrect overlap of the spray fans.
- Spray bar changing height as the distributor load decreases.
- Nozzles on spray bar not set at correct angle, the wrong size, different in size, plugged or restricted, or with imperfections.
- Inconsistent pump speed or pressure to the nozzles, and varying distributor travel speed.
- Emulsion allowed to cool below proper application temperature.
- Emulsion viscosity too high for existing conditions and equipment.

Bleeding

Bleeding or flushing is a surface that is too rich in asphalt (Figure 6.12). Bleeding can cause a slick, hazardous condition, especially during wet weather. The most common causes of bleeding of a chip seal are:

- Improper application rate of emulsion.
- Improper aggregate selection or application.
- Trapped moisture from the underlying pavement, resulting in stripping.
- Pre-existing bleeding of the underlying pavement.

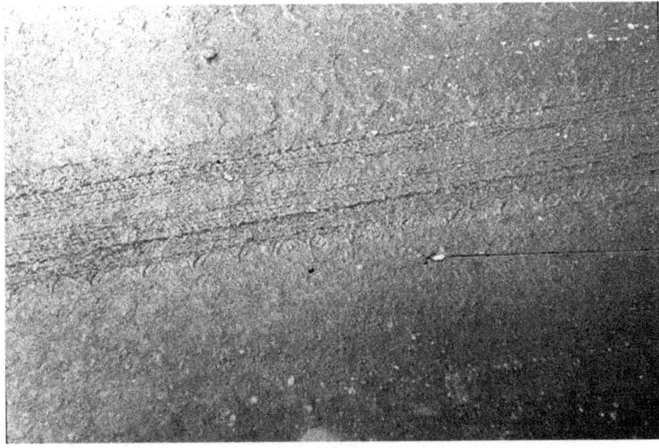

□ **Figure 6.12** Bleeding (Flushing)

Loss of Cover Aggregate

Loss of aggregate under traffic leaves the asphalt uncovered (Figure 6.13). This condition can be dangerous because loose aggregate thrown by the tires of a moving vehicle can cause windshield damage. Also, the aggregate-free surface can become a skid hazard.

Several causes of loss of cover aggregate include:

- Applying aggregate after the emulsion has broken.
- Using dirty aggregate.
- Excessive aggregate application.
- Rolling improperly.
- Applying too little emulsion.
- Improperly imbedding the aggregate.
- Applying while weather is too cool.
- Placing during high humidity.
- Placing on a surface which is too wet or dusty.
- Improperly controlling traffic.
- Having an absorptive surface that leaves too little asphalt in place to hold the aggregate.
- Washing away the emulsion if it rains during construction.
- Brooming too soon.

□ **Figure 6.13** Loss of Cover Aggregate

Sandwich Seal

A sandwich seal is constructed by spreading a large aggregate [5/8–3/4 in. (15–20 millimeters)], followed by the spraying of emulsion, and then an application of a smaller aggregate [1/4–1/2 inch (5–13 millimeters)] followed by normal chip seal rolling. The application of the large aggregate helps overcome the existing problem of a bleeding surface. The emulsion is normally a polymer-modified version of RS-2, CRS-2, or HFRS-2, and typically is applied at a rate greater than a single chip seal. The smaller aggregate locks down the larger aggregate. The aggregates must be clean and free of dust.

Sand Seal

A sand seal is an application of sprayed asphalt material covered with fine aggregate (chip seal using sand). It may be used to improve the skid resistance of slippery pavements or to seal against air and water infiltration. Construction procedures for sand seals are the same as those for single-chip seals. Usually, emulsion grades RS-1, CRS-1, MS-1, or HFMS-2 are used. They are applied at a rate of 0.15–0.20 gallons/square yard (0.70–0.90 liters/square meter). This is followed by 10–15 pounds/square yard (5.5–8 kilograms/square meter) of sand or screenings.

Dense-Graded Seal

Another type of surface treatment is the dense-graded seal. The graded aggregate seals are ideal for lightly travelled roads in sparsely populated areas where there is abundant gravel aggregate sources and traffic volumes are too low to justify the use of more expensive washed stone used in chip seals. The seal can be placed directly on a gravel surface or an existing paved surface without the requirement of priming first.

The graded aggregate used is typically all passing a 5/8-inch (16.0-millimeters) sieve, with from 45 to 65 percent passing a No. 4 (4.75 millimeters) sieve, and preferably having no more than 7 percent passing the No. 200 (75 microns) sieve. The asphalt emulsion grade used in the dense-graded seal is the medium-setting high-float type containing a small quantity of solvent. The use of the solvent helps the binder to penetrate through the dust coating and provide a good bond between the residual asphalt and the aggregate. A typical dense-graded seal would require a high-float emulsion application rate of 0.30 to 0.40 gallons per square yard (1.4 to 1.8 liter/square meter) to which from 30 to 40 pounds per square yard (16 to 22 kilogram/square meter) of graded cover aggregate would be applied.

Slurry Seal

A slurry seal is a mixture of well-graded fine aggregate, mineral filler (if needed), emulsified asphalt, and water applied to a pavement as a surface treatment, usually in a thickness of the largest aggregate size in the gradation. It is used in both the preventive and corrective maintenance of asphalt pavement surfaces. It does not increase the structural strength of a pavement section. Any pavement that is structurally weak in localized areas should be repaired before applying the slurry seal. All ruts, humps, low pavement edges, crown deficiencies, and waves should be corrected before placing a slurry seal.

A slurry seal can help reduce surface distress caused by oxidation of the asphalt surface. It will seal non-working cracks, stop raveling, make open surfaces less permeable to air and water and improve skid resistance and pavement appearance.

Slurry seal has a number of advantages, including:

- rapid application;
- no loose cover aggregate;
- excellent surface texture;
- the ability to correct minor surface irregularities;
- minimum loss of curb height;
- no need for manhole or other structure adjustments; and
- color contrast for lane delineation.

Materials

► Aggregates

The aggregate used in slurry seal must be clean, angular, durable, well-graded, and uniform. An individual aggregate or a blend of aggregates to be used in a slurry mix should meet these limits:

- Sand equivalent value, ASTM D 2419 (AASHTO T 176) = 45 minimum
- Los Angeles abrasion loss, ASTM C 131 (AASHTO 6) Grading C or D = 35 maximum

The three generally accepted gradations used for slurry mixtures are shown in Table 6.6.

Type I is used on lower volume roadways for maximum crack penetration. It is normally used in low-density traffic areas such as light-aircraft airfields, parking areas, or shoulders where the primary objective is sealing.

Type II is the most widely used gradation. It is used to seal, correct severe raveling and oxidation, and to improve skid resistance. It is typically used for moderate traffic.

Type III is typically used for arterial streets and higher volume roadways to correct surface conditions and to provide greater skid resistance.

☐ Table 6.6: Slurry Seal Aggregate Gradations*

Gradation Type	I	II	III
General Usage	Crack filling and fine seal	General seal, medium textured surfaces	Produces highly textured surfaces
Sieve Size	Percent Passing	Percent Passing	Percent Passing
3/8 in. (9.5 mm)	100	100	100
No. 4 (4.75 mm)	100	90–100	70–90
No. 8 (2.36 mm)	90–100	65–90	45–70
No. 16 (1.18 mm)	65–90	45–70	28–50
No. 30 (600 µm)	40–65	30–50	19–34
No. 50 (300 µm)	25–42	18–30	12–25
No. 100 (150 µm)	15–30	10–21	7–18
No. 200 (75 µm)	10–20	5–15	5–15
Residual Asphalt Content, % weight of dry aggregate	10–16	7.5–13.5	6.5–12
Application Rate, lb/yd^2 (kg/m^2), based on weight of dry aggregate	8–12 (3.6–5.4)	12–20 (5.4–9.1)	18–30 (8.2–13.6)

***Recommended by the International Slurry Surfacing Association.**

► Asphalt Emulsion

Emulsion used in the slurry mix may be SS-1h, CSS-1h, or CQS-1h.

► Other Materials

A small amount of additive may be added to the slurry mixture to control the breaking, set time, or cure time. A small amount of mineral filler such as hydrated lime, limestone dust, Portland cement, or fly ash may be necessary to aid in stabilizing and setting the slurry. Water used in the slurry should be potable and compatible with the mix.

▶ Mix Design

For slurry seal mix design, the following sources are recommended: ASTM D 3910 "Standard Practices for Design, Testing and Construction of Slurry Seal" or A-105 "Recommended Performance Guidelines for Emulsified Asphalt Slurry Seal Surfaces," available from the International Slurry Surfacing Association.

Blending the slurry seal materials in varying proportions in the laboratory is a great aid in selecting the proper mixture. Use the same materials in the laboratory design work that are intended to be used for the field construction. Correct blending should produce a slurry with a creamy texture that will flow smoothly in a rolling wave ahead of the strike-off squeegee. The slurry should be a semifluid, homogeneous mass with no liquid runoff.

▶ Construction Process

Slurry Seal Machine

The machine used for production of the slurry seal is a self-contained, continuous-flow mixing unit (see Figure 6.14). It is capable of delivering a predetermined amount of aggregate, mineral filler or other additive, and emulsion to the mixing chamber. Water is added to control consistency of the slurry mixture for proper application. The machine discharges the thoroughly mixed materials onto the prepared surface. Certain basic features are common to most slurry machines: truck-mounted units with separate storage tanks, bins, and metering systems for emulsion, water, aggregate, and mineral filler. The slurry machine has either a ribbon mixer or double shafted pugmill, from which the slurry is discharged into a spreader box. The spreader box is equipped with flexible squeegees, some with a device for adjustable placement width. Spreader boxes may be equipped with hydraulically powered augers to keep the slurry in motion and help keep the mixture uniformly spread across the spreader box width. The augers are helpful when quick-setting emulsion is used. One type of slurry mixer unit is depicted in the schematic drawing (Figure 6.15). Continuously self-loading machines are capable of placing many miles (kilometers) per day of slurry.

○ **Figure 6.14** Slurry Seal Equipment

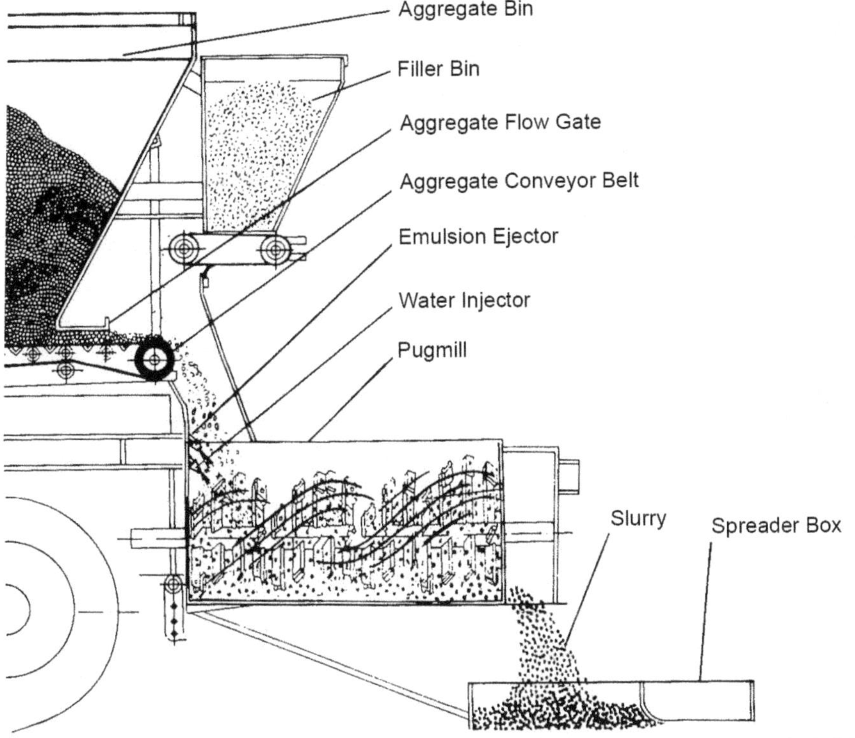

○ **Figure 6.15** Slurry Mixer Unit Schematic

Placing the Mixture

After mix proportions have been determined in the laboratory, it is advisable to place one or more trial mixes. The trial mixes serve a two-fold purpose:

- To calibrate the feeding and metering devices on the slurry machine. Aggregate flow should be determined for different gate openings and the amount of emulsion pumped per revolution of the aggregate feed belt.
- To ensure that the slurry mix characteristics are appropriate for field application, it is often necessary to make several trial runs.

Just before applying the slurry, the pavement surface should be cleaned of all dirt, dust, mud spots, vegetation, and other foreign matter. A tack/bond coat is not usually applied for HMA pavements but can be required in specific applications, such as for concrete, brick, or stone surfaces. An emulsion—usually a diluted CSS or SS grade— is applied at a rate of 0.05–0.10 gallons/square yard (0.19–0.38 liters/square meter). The pavement surface should be damp but with no free water in front of the slurry machine.

It is especially important to get a homogenous mix, one that will produce a slurry with a creamy texture that will flow smoothly in a rolling wave inside the spreader box. A non-homogenous mixture will cause an asphalt-rich surface and many ensuing problems. If extensive lumping, balling, unmixed aggregate, or aggregate segregation is observed, the slurry operation should be suspended until the problem is resolved. Streaks, such as those caused by oversized aggregate, should be repaired at once with a hand squeegee.

Care must be taken with longitudinal and transverse joints to prevent excessive buildup of slurry. Common practice is to make certain the joint of the adjacent lane is either completely cured or is still in a semi-fluid condition. For good appearance and durability, a joint should not be made when the previous pass is only partially set because tearing and scarring might result.

A canvas or burlap drag may be pulled behind the spreader box to improve the joint and overall surface appearance. Drags should be changed regularly. Hand squeegees and hand drags are used to improve joints and place the slurry in areas inaccessible to the machine.

On high-crowned pavements or superelevated curves, slurry should be diverted to the high side of the spreader box. Gravity will keep the low side filled. Spreading slurry in hilly areas is easier if the slurry machine travels in an uphill direction. If circumstances require placement in a downhill direction, the slurry may be thickened to prevent it from flowing ahead of the machine.

Rolling

Rolling of slurry seals is usually not needed. However, it may be useful in those areas where it will improve short-term durability. Such areas include taxiways, runways, parking lots, truck terminal yards, and areas where there is insufficient traffic to consolidate the mixture. These areas are subject to power-steering turns, braking, or acceleration forces. When rolling, a 5-ton (4500-kilogram) pneumatic roller with 50 pounds per square inch (345 kiloPascals) tire pressure will be most effective. Rolling can start as soon as clear water can be pressed out of the slurry mixture with a piece of paper without discoloring the paper. In most cases, traffic will iron out the slurry and close any hairline cracks of dehydration.

Curing/Traffic Control

Slurry should be placed only when the pavement and air temperature are at least 50°F (10°C) and rising and when no rain is expected. Roads with newly placed slurry should not be opened to traffic until the slurry has cured sufficiently to support straight rolling traffic. As with rolling, traffic generally can be allowed on the slurry as soon as clear water can be pressed out of the slurry mixture with a piece of paper without discoloring the paper. The traffic must be controlled because quick stops or accelerations and the turning of a parked vehicle's wheels will damage the newly placed slurry.

Micro Surfacing

Micro surfacing is similar to slurry seal but utilizes a polymer-modified binder and higher quality aggregate. A polymer-modified binder makes a much stiffer mix which allows the mixture to be placed more than one aggregate in thickness. Because of the mix stiffness, a more substantial mixer is required, and twin-shafted paddles or spiral augers are required to provide a uniform flow of the mix in the spreader box.

Micro surfacing consists of a polymer-modified emulsion, a high-quality aggregate, mineral filler, additives, and water. A typical micro surfacing mixture includes 82–90 percent aggregate and the following as a percentage of the dry aggregate:

- 0.5–3.0 percent mineral filler, and
- 5.5–9.5 percent residual asphalt binder.

Materials

The emulsion is a quick-setting, polymer-modified emulsion. Aggregates used for micro surfacing are manufactured, 100 percent crushed stone such as granite, slag, limestone, chat, or other high-quality aggregate. An individual aggregate or blend of aggregates must meet these standards for use in micro surfacing:

- Sand equivalent value, ASTM D 2419 (AASHTO T 176) = 60 percent minimum.
- Soundness, ASTM C 88 (AASHTO T 104) = 15 percent maximum using Na_2SO_4 or 25 percent maximum using $MgSO_4$.
- Los Angeles abrasion loss, ASTM C 131 (AASHTO T 96) Grading C or D = 30 percent maximum.

The aggregates gradations used for micro surfacing should meet the requirements of Type II and Type III aggregates as set forth in Table 6.7.

☐ **Table 6.7: Micro Surfacing Aggregate Gradations***

Gradation Type	II	III
General Usage	General resurfacing, sealing and renewal of surface friction	High volume roadway resurfacing, rut filling. Produces high-friction surfaces
Sieve Size	Percent Passing	Percent Passing
3/8 in. (9.5 mm)	100	100
No. 4 (4.75 mm)	90–100	70–90
No. 8 (2.36 mm)	65–90	45–70
No. 16 (1.18 mm)	45–70	28–50
No. 30 (600 µm)	30–50	19–34
No. 50 (300 µm)	18–30	12–25
No. 100 (150 µm)	10–21	7–18
No. 200 (75 µm)	5–15	5–15
Residual Asphalt Content, % weight of dry aggregate	5.5–9.5	5.5–9.5
Application Rate, lb/yd^2 (kg/m^2) based on weight of dry aggregate	12–20 (5.4–9.1)	18–30 (8.2–13.6)

***Recommended by the International Slurry Surfacing Association.**

The most common types of mineral filler are non-air entrained Portland cement or hydrated lime that is free from lumps. Other types of mineral fillers have been used successfully. The type and amount of mineral filler is determined by a laboratory mix design. Adjustments to the mineral filler content, consistent with the mix design tolerances, may be required when the micro surfacing is being placed to provide better consistency or set times.

A small amount of additive may be added to the micro surfacing mixture to control the breaking, set time, or cure time. Water used in the micro surfacing mixture should be potable and compatible with the mix.

Materials evaluation and mix design should be done in accordance with the International Slurry Surfacing Association's *"Recommended Performance Guide for Micro-Surfacing, A143"* (latest version).

▶ Equipment

Mixing Equipment

The micro surfacing is mixed in a continuous-flow mixing unit (see Figure 6.16) that can accurately deliver and proportion the aggregate, emulsion, mineral filler, control-setting additive, and water to a revolving multi-blade, double-shafted mixer and to discharge the mixed micro surfacing on a continuous-flow basis. The micro surfacing mixing unit may be required to be a self-loading machine capable of loading material while continuing to lay micro surfacing, thereby minimizing construction joints.

Spreading Equipment

The mixture is spread uniformly in the surfacing box (see Figure 6.16) by means of paddles or augers inside the spreader box. A front seal is provided to ensure no loss of the mixture at the road contact point. A rear seal acts as a final strike-off. The spreader box and rear strike-off are designed and operated to achieve a uniform consistency and to produce a free flow of material to the rear strike-off. One feature of micro surfacing is the ability to fill minor depressions such as ruts, utility cuts, or channels in the traffic wheelpaths, provided the pavement is no longer in a plastic flow condition. It can be used to fill ruts, utility cuts, depressions in the existing surface, etc. Ruts

of 1/2 inch (13 millimeters) or greater in depth can be filled independently with a rut-filling spreader box (see Figure 6.17) either 5 feet (1.5 meters) or 6 feet (1.8 meters) in width. For irregular or shallow rutting of less than 1/2 inch (12.5 millimeters) in depth, a full-width scratch-coat pass may be used. Ruts that exceed a depth of 1 1/2 inches (37.5 millimeters) may require multiple placements with the rut-filling spreader box to restore the cross-section. All rut-filling material should cure under traffic for at least 24 hours before additional material is placed.

▶ Weather Limitations

Micro surfacing should not be applied if either the pavement or air temperature is below 50°F (10°C) and falling, but application may start when both pavement and air temperatures are above 45°F (7°C) and rising. No micro surfacing should be applied when the possibility exists that the finished product will freeze within 24 hours.

▶ Surface Preparation

Before micro surfacing is placed, the surface of the existing pavement should be cleaned, prepared, and free of standing water. The pavement should be thoroughly broomed and cleaned to remove any loose dirt and contaminants. Manholes, valve boxes, drop inlets, and other service entrances shall be protected from the micro surfacing by a suitable method. Cracks greater than 1/4-inch wide should be sealed prior to application.

A tack coat is not usually applied for HMA pavements but can be required in specific applications, such as for concrete, brick, or stone surfaces. An emulsion—usually a diluted CSS or SS grade—is applied at a rate of 0.05–0.10 gallons/square yard (0.19–0.38 liters/square meter).

▶ Application

It is recommended that a test strip be placed in conditions similar to those expected during the project. Pre-wetting the surface by applying a light water fog ahead of the spreader box is effective when the weather is hot or the pavement texture is rough. The water keeps the emulsion from prematurely breaking before the mix can be spread and aids in the filling of small surface cracks.

▶ Joints

No excess buildup, uncovered areas, or unsightly appearance should be permitted on longitudinal or transverse joints. The goal is to produce a minimum number of longitudinal joints throughout the project. When possible, longitudinal joints should be placed on lane lines. Partial width passes should be used only when necessary. If partial width passes are used, they should not be the last passes of any paved area. A maximum of 3 inches (75 millimeters) is recommended for the overlap of longitudinal lane line joints. Also, the joint should have no more than a 1/4-inch (6-millimeter) difference in elevation when measured by placing a 10-foot (3-meter) straightedge over the joint and measuring the elevation drop-off.

Cape Seal

The term *Cape seal* is attributed to the Cape Provincial Administration of South Africa. A Cape seal can be defined as a single-layer chip seal followed by a slurry seal. In order to have a successful Cape seal project, it is important to follow standard chip seal/slurry seal specifications and construction methods. A minimum cure time of three days between placement of the chip seal and subsequent slurry seal application is recommended. In order to achieve the desired surface texture of a Cape seal, it is important to avoid an excess of slurry. Two typical Cape seal designs are shown in Table 6.8.

☐ **Table 6.8: Quantities of Emulsion and Aggregate for Cape Seal**

	Nominal Aggregate Size	Size No.	Aggregate Quantity lb/yd^2 (kg/m^2)	Emulsion Quantity gal/yd^2 (L/m^2)	Slurry Rate (Type I) lbs/yd^2 (kg/m^2)
1/2 in. (12.5 mm) thick	3/8 inch, to No. 8 (9.5 to 2.36 mm)	7	25–30 (14–16)	0.30–0.45 (1.4–2.0)	6–10 (2.7–4.5)
3/4 in. (19.0 mm) thick	3/4 to 3/8 inch (19.0 to 8.5 mm)	6	40–50 (22–27)	0.40–0.50 (1.8–2.3)	8–12 (3.5–5.5)

CHAPTER 7

EMULSION-AGGREGATE MIXES

The scope of this chapter is asphalt emulsion mixtures composed of virgin aggregates, existing subgrade or unbound aggregate layers, stabilized with asphalt emulsions. Chapter 8 will discuss asphalt emulsion mixtures composed of recycled asphalt materials stabilized with asphalt emulsions.

Advances in asphalt emulsion technology make it possible for emulsion mixes to be used in a wide variety of pavement construction, rehabilitation, and maintenance applications. Table 7.1 lists the major uses of asphalt emulsion mixes.

☐ Table 7.1: Major Uses of Asphalt Emulsion Mixtures

Mixture Use	Purpose of Emulsion Treatment
As a construction aid	To facilitate the construction of the pavement and, in some cases, provide a working platform.
Upgrading of marginal aggregates	To improve an aggregate to the quality of a good untreated granular base.
As a temporary wearing surface	To provide a surface that may be used until a permanent surface of hot mix asphalt or high-type emulsion mix is placed.
To reduce the total pavement thickness	To increase the strength of the pavement materials and reduce the required pavement structure thickness from that required using untreated materials.
Base and wearing surface	To produce a high-quality mix for all traffic load applications. These mixes have good flexibility and resistance to permanent deformation.
Pavement subbase	To allow the use of local or in-place materials. Sands, silty sands, and poorer graded sand-gravels can be used to produce an acceptable subbase layer.
Maintenance mixtures	To provide workable mixtures designed for immediate use or long-term storage.

Asphalt Emulsion Mixtures for Base and Wearing Surfaces

There are three types of asphalt emulsion-aggregate mixtures: dense-graded, sand, and open-graded. The aggregate properties for each of these mix types are shown in Tables 7.2, 7.3, and 7.4. The emulsion types used are the slow-setting or medium-setting grades.

Dense-graded mixtures have aggregates that are graded from the maximum size down to and including material passing the No. 200 (75 µm) sieve. They include a wide variety of aggregate types and gradations and can be used for all types of pavement applications. A wide range of emulsion grades are used based on environment and traffic conditions.

Sand emulsion mixtures are produced by treating bank-run sands, poorly graded sand-gravels, and "dune" or "sugar" sands with asphalt emulsions. Sand mixes are generally restricted to fine granular sands and silty sands low in clay content. Sand mixes have provided good performance as subbase and base layers when produced with the proper emulsion grades. Since these sand mixes lack aggregate interlock, a harder or "h" grade (i.e., CSS-1h) is preferred to enhance mixture stability.

Open-graded mixtures provide high air voids to drain water through the mix. These mixtures have been used successfully for both base and surface courses. These mixes have shown good resistance to fatigue, reflection cracking, rutting, and shoving. Due to the lack of fines in these mixtures, a higher viscosity or "2" grade (i.e., CMS-2) is preferred to increase film thickness for better durability.

Dense-graded and sand mixes normally require a surface seal for waterproofing and to provide a more durable wearing surface. Chip seals are frequently used for dense-graded mixes, but a fog or sand seal may be adequate in some situations. For open-graded mixes, a surface seal is not typically required. As with any emulsion surface treatment, surface seals for asphalt emulsion-aggregate mixes need to be properly designed and applied. Do not apply the surface seals until the mixture has completely cured.

Mix Design Methods

A mix design is required for asphalt emulsion-aggregate mixes. It is essential that trial mixes be prepared in the laboratory to determine the grade and percent of emulsion and mixture properties of workability, stability, and strength. The emulsion mixture's susceptibility to water damage should be determined.

The decision on selecting a type and grade of emulsion should consider the characteristics of both the aggregate (gradation, cleanliness, and mineralogy) and the asphalt emulsion (hard or soft base, containing solvent, polymer-modified, and cure rate).

There is no universally accepted asphalt emulsion-aggregate mix design method for either dense- or open-graded cold mixtures. However, nearly all of the dense-graded methods are modifications of the Marshall (ASTM D 6926 and 6927 or AASHTO T 245) or Superpave (ASTM D 7229) design methods.

This section standardizes two design methods for use with asphalt emulsion cold mixtures. One method is for the design of mixes having dense-graded aggregate and the second for mixes having open-graded aggregate. These procedures may be used as guides for developing mix design method templates reflecting local materials and conditions.

Dense-Graded Aggregate Cold Mix Design
This design method is for asphalt emulsion cold mixtures containing dense-graded mineral aggregates with a maximum size of 1 inch (25 millimeters) or less and either medium or slow-setting types of emulsions. This design method is applicable to mixtures produced by either in-place or plant mixing at ambient temperatures and whether placed immediately or stockpiled for later placement.

Aggregates

Dense-graded aggregates meeting the requirements of Table 7.2 are among those suitable for asphalt emulsion mixtures. For the gradations containing appreciable fines, aeration or drying prior to compaction may be required.

Asphalt Emulsions

Two types of asphalt emulsions are used for producing dense-graded emulsion cold mixtures—slow-setting (SS) and medium-setting (MS) types listed in Table 5.1. Medium-setting emulsions are normally used with aggregates that do not have excessive amounts of material passing the No. 200 (75 microns) sieve and/or if stockpiling of the mixture is desired. Conversely, slow-setting emulsions normally are used with more dense-graded aggregates [higher amounts of material passing No. 200 (75 microns) sieve] and stockpiling is not desired.

Determination of Trial Emulsion Content

There are several procedures available to determine a design starting point for the trial emulsion or residual asphalt content of a mixture. For this design procedure, two formulas are used within the procedure—one for base mixtures and another one for surface mixtures. The formulas are based on the percentage of aggregate passing the No. 4 (4.75 millimeters) sieve and in most cases will give a satisfactory starting point.

1. Determine the residue content of the asphalt emulsion using ASTM D 6997, "Residue and Oil Distillate by Distillation."
2. Estimate an initial emulsion content based on the dry weight of aggregate using the appropriate formula for the mix being designed:

$$\text{Base Mixtures} \qquad \text{Percent Emulsion} = \frac{[(0.06 \times B) + (0.01 \times C)] \times 100}{A}$$

$$\text{Surface Mixtures} \qquad \text{Percent Emulsion} = \frac{[(0.07 \times B) + (0.03 \times C)] \times 100}{A}$$

where:

Percent Emulsion = Estimated initial percent asphalt emulsion by dry weight of aggregate

A = Percent residue of emulsion by distillation (Step 1)

B = Percent of dry aggregate passing No. 4 (4.75 millimeters) sieve

$C = 100 - B$ = Dry aggregate retained on No. 4 (4.75 millimeters) sieve

Coating and Adhesion (Stripping) Testing

The preliminary evaluation of each asphalt emulsion selected for mixture design is accomplished through a coating test and an adhesion test. The trial emulsion content determined above is combined with the pre-wetted job aggregate, corrected to a dry weight. The coating is visually estimated as satisfactory or unsatisfactory for the intended use of the mix. Surface mixtures normally require a greater degree of coating than do base mixtures.

Asphalt emulsion-aggregate mixtures require some amount of pre-mixing water. If premature breaking of emulsion around aggregate fines occurs (balling), mixing at additional water contents should be evaluated. If the degree of coating is considered satisfactory, then the adhesion test is completed. If the coating is considered unacceptable, the type or formulation of the emulsion used may be changed and the mix design repeated.

Coating Testing Procedure

(1) Determine the moisture content of a representative sample of the aggregate. Care must be taken to maintain the moisture in the field sample. If the aggregate is received dry or dried for blending, the estimated stockpile moisture must be added to the aggregate or individual combined aggregate samples 24 hours prior to performing any test.

(2) Weigh the equivalent of 500 grams of dry aggregate (500 grams + moisture) into a suitable mixing bowl.

(3) If required, weigh in the premixing water and mix by hand for 10 seconds or until visually uniformly dispersed.

(4) Weigh in the trial emulsion content to the pre-wetted aggregate at the anticipated use temperature and mix vigorously by hand for a maximum of 60 seconds or until sufficient dispersion has occurred throughout the mixture.

(5) Place the mixture on a flat light-colored surface and visually estimate the degree of coating.

(6) The cured emulsion-aggregate mixture may be evaluated for water resistance. Totally submerse the mix in water (about twice the volume of water to mix) and then pour off the water. Place the mixture on a flat light-colored surface and visually estimate the degree of retained coating. If the coating is not acceptable, then the type or formulation of the emulsion used may be changed.

Adhesion (Stripping) Testing Procedure

(1) Cure a 100-gram portion of the above produced mix in a shallow container for 24 hours in a forced draft oven at 140°F (60°C).

(2) Put the oven-cured mix in a 600-milliliter beaker containing 400 milliliters of boiling distilled water.

(3) Bring the water back to a boil and stir the water at one revolution per second for three minutes.

(4) Pour off the water and place the mix on a piece of light-colored absorbent paper.

(5) After the mix has dried, visually evaluate the amount (percent) of retained asphalt coating. If satisfactory, continue the mix design. If coating is not acceptable, then the type, grade, or formulation of the emulsion used may be changed.

Preparation of Test Specimens

Prepare three or more specimens at a minimum of three different emulsion contents, with one below and one above the trial emulsion content. If the mixture in the coating test appears to be dry, start with the trial emulsion content and increase the content for each of the other two content levels. Conversely, if the mixture in the coating test appears rich, reduce the content for the other levels. A normal difference between the emulsion contents is one percent, or a residual asphalt content difference of 0.65 percent for an emulsion with a 65 percent residual content.

Mixing Procedure

(1) Weigh into suitable mixing bowls the appropriate amount of pre-wetted job aggregate to obtain a compacted specimen height of 2.5 ± 0.25 inches (63.5 ± 6 millimeters) for each individual batch. The amount normally required is about 1200 grams of dry aggregate. Care must be taken so that the aggregate for each batch is representative of the project aggregate. If necessary, the aggregate can be dried, separated into sizes, and then reblended into individual batch sizes. If this is done, water equivalent to the stockpile or desired moisture content must be added to each batch and the batch covered to prevent loss of moisture for about 24 hours prior to mixing with emulsion.

(2)   If pre-wetting water is required, weigh onto the aggregate the amount as determined in the coating test and hand mix for 10 seconds or until the moisture is uniformly dispersed. This must be completed immediately prior to the addition and mixing of the emulsion.

(3)   Weigh the predetermined amount of emulsion onto the ambient pre-wetted aggregate and stir vigorously for a maximum of 60 seconds or until sufficient emulsion dispersion has occurred.

Compaction Procedure

(1)   Aeration or drying of a dense-graded mixture is often required prior to specimen compaction. Anytime the total liquid volume (emulsion + water in aggregate) exceeds the voids in mineral aggregate (VMA) plus any absorbed liquid volume, proper compaction cannot be achieved. This condition can be identified if the Marshall hammer bounces and/or liquid exudes from the specimen. When this condition exists, place the mixture in a shallow container and use a fan and occasional stirring to reduce the moisture content so that proper compaction can be achieved. Specimens that could not be compacted satisfactorily should be discarded and not retested.

(2)   Thoroughly clean the specimen mold assembly and the face of the compaction hammer. Place a paper disc in the bottom of the mold before the mixture is introduced. Transfer the entire batch into the mold and spade the mixture vigorously with a spatula 15 times around the perimeter and 10 times over the interior of the mold. With the spatula, smooth the surface of the mix into a slightly rounded shape.

(3)   Place the mold assembly on the compaction pedestal in the mold holder and apply 50 blows with the compaction hammer with a free fall of 18.0 inches (457 millimeters). Remove the base plate and collar and reverse the molded specimen and reassemble. Apply 50 blows of compaction to the face of the reversed specimen.

(4)   Remove the base plate, collar, and paper discs, and place the mold containing the compacted specimen on a perforated shelf in a 140°F (60°C) forced draft oven for 48 hours. Ensure the specimen is at the bottom of the mold so that the oven shelf supports it during curing.

(5)   Remove the mold containing the compacted specimen from the oven and while still at 140°F (60°C), apply a static load of 40,000 pounds (178 kiloNewtons) by the double plunger method where a free-fitting plunger is placed at both the bottom and top of the specimen in the mold. Apply the load at a rate to give about 0.05 inches/minute (1.3 millimeters/minute) of compression and maintain the full load for one minute and then release.

(6)   Allow the compacted specimen to cool in the mold for a minimum of one hour prior to extracting the specimen for testing.

Testing of Compacted Specimens

Approximate volumetrics and stability values can be determined from the compacted specimens if desired. Often, the volumetrics are not evaluated because they are normally calculated as approximations because of the possibility of some moisture in the compacted specimens. If more exact values are desired, the moisture must be accounted for in the compacted specimens and the theoretical maximum density must be determined using moisture free, loose mixture.

Volumetrics

(1) The easiest and most often used method to determine bulk density is to divide the weight of the specimen in air by its volume. The bulk density may be determined so that the compaction and/or composition of like specimens is validated.

$$D_b = \frac{W_a}{HA}$$

where:

D_b = Measured bulk density of a compacted mixture specimen
W_a = Compacted specimen weight in air
H = Height of compacted specimen
A = Cross sectional area of a compacted specimen (πr^2)

(2) Other volumetrics, such as voids, voids filled, and voids in mineral aggregate, may be determined by properly accounting for moisture and following the appropriate ASTM testing procedures, including D 70, D 1188, D 2041, D 2726, and D 3203.

Stability Testing

(1) Marshall stability and flow are determined following the procedures of ASTM D 6927, except that the compacted specimens shall be placed in an air bath for a minimum of two hours at the test temperature of $77 \pm 1.8°F$ ($25 \pm 1°C$). A stability value of 500 pounds (2224 Newtons) or greater has been found to be satisfactory for most pavements with low to moderate traffic volumes. Local experience may justify a different minimum stability value.

Open-Graded Aggregate Cold Mix Design

This mix design method covers procedures for preparing trial mixtures of open-graded emulsion mixtures using aggregates having properties as indicated in Table 7.3. Job aggregate and an asphalt emulsion are combined to establish a design emulsion content based on an evaluation of runoff. Medium-setting asphalt emulsions are typically used for open-graded mixtures.

To summarize this method, mixtures are made with varying emulsion contents in one percent increments and subjected to a runoff method. The asphalt emulsion content that gives a runoff of 10 grams of asphalt residue is recommended as the design emulsion content.

Preparation of Mixtures

These procedures are used for the preparation of open-graded aggregate-asphalt emulsion mixtures for testing and determination of the optimum emulsion content.

a. Obtain a representative sample of job aggregate and dry to a constant weight in a forced draft oven at 140°F (60°C). After oven drying, cool the sample at ambient temperature for a minimum of two hours.

b. Preweigh a sufficient number of 2000-gram batch samples of the dried aggregate using the stockpile gradation.

c. Using suitable mixing bowls, mix the aggregate with 40 grams of water until all of the aggregate is damp. Cover the bowl with a clean cloth and leave covered for 15 minutes.

d. Add the appropriate amount of asphalt emulsion preheated to 140°F (60°C) to the dampened aggregate and hand mix for 2 minutes. Observe and record the workability of the mix, such as stiff, satisfactory, or loose, and the percent of coating. Emulsion contents in 1.0 percent increments by weight of dry aggregate are recommended. A beginning emulsion content of 4.0 percent is recommended for very coarse maximum size aggregates and 6.0 percent for the finest maximum size aggregates.

Testing Procedures

a. Lightly dampen a No. 8 (2.36 millimeters) sieve with water and place on a tared standard pan. Immediately after preparation of the mixture, transfer the entire batch onto the No. 8 sieve.

b. Allow the mix to drain into the standard pan at ambient temperature for 30 minutes.

c. Remove the drained mix from the No. 8 sieve and spread the mix onto a paper (light colored) lined tray. Surface dry the mix with a fan and evaluate the percent of asphalt coating.

d. Place the standard pan containing runoff in a forced draft oven at 230 ± 9°F (110 ± 5°C) and dry to a constant mass. Weigh the pan containing the runoff, subtract the tared weight of the pan, and record this as asphalt residue runoff in grams (W).

Optimum Emulsion Content Determination

a. Plot the percent Emulsion Content by Weight of Aggregate versus the Asphalt

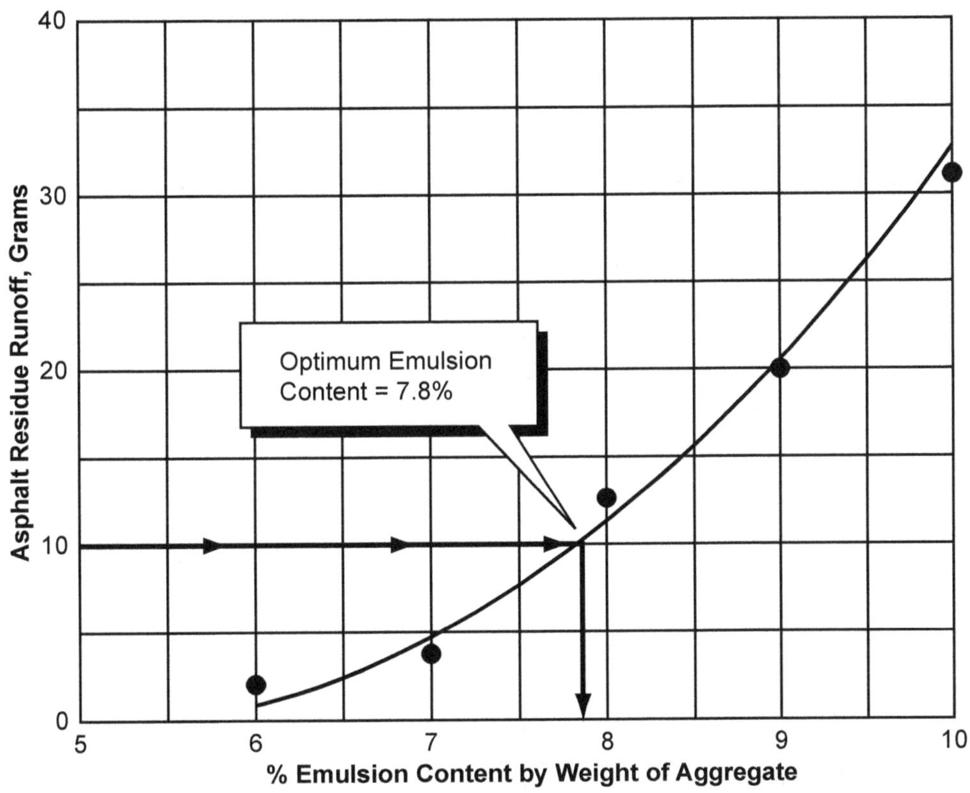

☐ **Figure 7.1**   Optimum Emulsion Content Plot

Residue Runoff (W) on a graph and draw a smooth curve (Figure 7.1).

b. Draw a horizontal line at 10 grams of Asphalt Residue Runoff on the *Y*-axis to intersect with the curve. Read the corresponding emulsion content on the *X*-axis to the nearest 0.1 percent. This is the optimum emulsion content.

Final Results

a. Report these results:

    (1) Optimum Emulsion Content, percent

    (2) Mix Workability (stiff, satisfactory, or loose)

    (3) Mix Coating, percent

At the optimum asphalt emulsion content, the mixture must have satisfactory workability. It is preferred that coating be as close as possible to 100 percent. Mixes shall be considered suitable, however, if they have a minimum of 85 percent coating if used as a surface course and 60 percent coating if used as a base course.

Aggregates

Aggregate makes up 90 to 95 percent by weight of an emulsion mixture. The characteristics of the aggregate are very important in obtaining desired physical properties and good field performance.

A wide variety of aggregate types and gradations may be used successfully for emulsion mixes. Tables 7.2, 7.3, and 7.4 show some of the typical gradations and other aggregate properties for dense-graded, open-graded, and sand emulsion mixes, respectively. Certain standards must be maintained for the quality of the aggregates, including the amount of material finer than No. 200 (75 microns) sieve, the plastic fines content, and durability. The test methods listed in Table 7.5 are those typically used to evaluate the aggregate physical properties for emulsion mixes. Compatibility of the aggregate with the asphalt emulsion is important and should be determined. The mineral composition of the aggregate can have a significant influence on the performance of the mix. Therefore, as previously noted, it is necessary to prepare trial mixtures in the laboratory.

☐ **Table 7.2: Aggregates for Dense-Graded Emulsion Mixtures**

Sieve Size	Semi-Processed-Crusher, Pit or Bank Run	Percent Passing by Weight				
		Processed Dense-Graded Asphalt Mixtures				
2 in. (50.0 mm)	–	100	–	–	–	–
1 1/2 in. (37.5 mm)	100	90–100	100	–	–	–
1 in. (25.0 mm)	80–90	–	90–100	100	–	–
3/4 in. (19.0 mm)	–	60–80	–	90–100	100	–
1/2 in. (12.5 mm)	–	–	60–80	–	90–100	100
3/8 in. (9.5 mm)	–		–	60–80	-	90–100
No. 4 (4.75 mm)	25–85	20–55	25–60	35–65	45–70	60–80
No. 8 (2.36 mm)	–	10–40	15–45	20–50	25–55	35–65
No. 16 (1.18 mm)	–	–	–	–	–	–
No. 30 (600 µm)	–	–	–	–	–	–
No. 50 (300 µm)		2–16	3–18	3–20	5–20	6–25
No. 100 (150 µm)	–			–	–	–
No. 200 (75 µm)	3–15	0–5	1–7	2–8	2–9	2–10
Sand Equivalent Percent	30 min.	35 min.	35 min.	35 min.	35 min.	35 min.
LA Abrasion @ 500 Revolutions	–	40 max.	40 max.	40 max.	40 max.	40 max.
Percent Crushed Faces	–	65 min.	65 min.	65 min.	65 min.	65 min.

☐ **Table 7.3: Aggregates for Sand Emulsion Mixtures**

Sieve Size	Total Percent Passing		
	Poorly-Graded	Well-Graded	Silty Sands
1/2 in. (12.5 mm)	100	100	100
No. 4 (4.75 mm)	75–100	75–100	75–100
No. 50 (300 µm)	–	15–30	–
No. 100 (150 µm)	–	–	15–65
No. 200 (75 µm)	0–12	5–12	12–20
Sand Equivalent, percent	40 max.	40 max.	40 max.
Plasticity Index	65 min.	65 min.	65 min.

☐ Table 7.4: Aggregates for Open-Graded Emulsion Mixtures

Sieve Size	Base			Surface Course
	Coarse	Medium	Fine	
1 1/2 in. (37.5 mm)	100	–	–	–
1 in. (25.0 mm)	95–100	100	–	–
3/4 in. (19.0 mm)	–	90–100	–	–
1/2 in. (12.5 mm)	25–60	–	100	–
3/8 in. (9.5 mm)	–	20–55	85–100	100
No. 4 (4.75 mm)	0–10	0–10	–	30–50
No. 8 (2.36 mm)	0–5	0–5	0–10	5–15
No. 16 (1.18 mm)	–	–	0–5	–
No. 200 (75 μm)	0–2	0–2	0–2	0–2
LA Abrasion @ 500 Revolutions	40 max.	40 max.	40 max.	40 max.
Percent Crushed Faces	65 min.	65 min.	65 min.	65 min.

☐ Table 7.5: Aggregate Physical Properties

Characteristics	Method of Test	
	ASTM	AASHTO
Amount of material finer than No. 200 (75 μm)	C 117	T 11
Unit weight of aggregate	C 29	T 19
Sieve analysis, fine and coarse aggregates	C 136	T 27
Sieve analysis of mineral filler	D 546	T 37
Toughness of coarse aggregates (LA Abrasion)	C 131	T 96
Plastic fines in graded aggregates and soils (Sand Equivalent Test)	D 2419	T 176
Freeze Thaw Durability (Sulfate Soundness Test)	C 88	T 104
Reactivity and Absorption (Methylene Blue)	C 837	TP 57

Additives

The most common additives used in emulsion mixes are Portland cement and hydrated lime. These additives can be very effective in obtaining higher early strength and reducing the water susceptibility of emulsion mixes, particularly those produced with sand and sand-gravel aggregates. The amount of cement and lime used is typically 1 to 2 percent by weight of dry aggregate. These materials may be added dry or in a slurry. Laboratory testing is required to determine if the additives are of sufficient benefit to justify the increased cost.

Production of Emulsion-Aggregate Mixtures

Two methods are used to produce emulsion-aggregate mixtures: (1) mixed-in-place, or (2) central plant mix.

There are a number of factors to be considered when selecting the type of emulsion-aggregate mixture production method. The factors to be considered in determining the most appropriate mixing method should include: (1) project location, (2) project size, (3) traffic conditions to maintain, or if the road can be closed, (4) whether imported aggregate is necessary to improve the mixture properties, (5) pavement type and thickness, and (6) climatic conditions.

Mixed-in-Place Process

The mixed-in-place process can use a variety of mixing procedures, such as a motor-grader blade mixing, reclaimer machine mixing, or travel plant mixing. All of these in-place-mixing processes will stabilize the existing roadbed. A laboratory evaluation will determine if the addition of an imported aggregate will be necessary to improve the mixture properties. If additional aggregate is used, it is placed on the surface of the material to be stabilized. The amounts of new aggregate and emulsion are carefully controlled to ensure the proper proportions necessary to meet the target mixture properties.

▶ Blade Mixing

For blade mixing, the asphalt emulsion is applied by an asphalt distributor on a flattened windrow of in-place and/or imported material immediately ahead of the motor grader. The blade of the motor grader mixes the materials through a series of tumbling and rolling actions (Figure 7.2). If water is required to aid in mixing and compaction, it is applied to the material prior to the addition of emulsion. To assure a more uniform windrow, place the windrow material with a spreader box or windrow sizer. The emulsion required by the material in the windrow must be calculated and controlled.

◻ **Figure 7.2**　Blade Mixing Operations

Blade mixing is the least precise of the mixed-in-place methods and is primarily applicable when short lengths or small areas are being stabilized. The success of blade mixing is very dependent on the experience and capability of the motor-grader operator. When blade mixing, the mold board of the motor grader should give a rolling action to the material. Care must be used to prevent extra material from being taken from the roadway and incorporating it into the mixture being processed.

There is a possibility of variation in the gradation of the material in the windrow, requiring a fluctuation in the emulsion demand. Therefore, very close attention must be given to the appearance of the mix as mixing progresses. It is important that as much uniformity as possible be obtained in the material gradation and emulsion and moisture contents. The mixing operation should continue until adequate coating is achieved. However, excessive mixing should be avoided as it can lead to stripping of the asphalt coating from the aggregate. After the mixing has been completed, the material should be moved to one side of the roadbed in preparation for spreading.

▶ Rotary/Reclaimer Mixing

Rotary mixers (Figure 7.3) have been used for many years for the in-place mixing of asphalt emulsions. These mixers consist of a mobile mixing chamber mounted on a self-propelled machine. The chamber is open on the bottom with a transverse rotating shaft equipped with tines or cutting blades.

☐ **Figure 7.3** Rotary Mixer

Reclaimers (Figure 7.4) are heavier, more powerful machines equipped with carbide-tipped teeth for more effective pulverization of existing asphalt and other road materials. This equipment is increasingly being used instead of rotary mixers for the in-place mixing of emulsions.

☐ **Figure 7.4** Reclaimer Machine

Some rotary mixers and reclaimers have liquid additive systems with a spray bar inside the mixing chamber available for both emulsion and water. For machines without liquid additive systems, multiple passes will be required adding water and emulsion between passes.

► **Travel Plant Mixing**

Travel plants (Figure 7.5) are self-propelled pugmill mixing plants that proportion and mix asphalt emulsion and aggregates in place as they move along the road. The travel plant receives aggregate into a hopper from a haul truck, adds and mixes the asphalt emulsion in a pugmill, and spreads the mix at the rear by a screed as it moves forward on the surface being paved. The travel plant has a tank for the storage of asphalt emulsion or can be supplied from an emulsion tanker.

The purpose of the travel plant is to produce a uniform, properly coated, asphalt emulsion-aggregate mixture. The travel plant should have the means of proportioning the amount of aggregate and emulsion to the mixing chamber.

☐ **Figure 7.5**   Travel Plant

► **Placement and Compaction of In-Place Mixtures**

The mixture should always be spread to a uniform thickness, whether in a single pass or in several thinner layers, so that no thin areas exist in the final mat. No layer should be thinner than about two times the maximum dimension of the largest aggregate. As a general rule, maximum layer thickness should not exceed six times the maximum dimension of the largest aggregate in order to achieve proper compaction. Also, experience has shown sand mixes should be placed in compacted thicknesses of no greater than 2 inches (50 millimeters).

Breakdown or initial rolling of emulsion mixes should begin when its final required cross section is obtained and just before or as the emulsion begins to break. At this point the mixture should be able to support the roller without excessive displacement. If the mixture ruts or shoves during compacting, rolling should be discontinued until there is a reduction in the moisture content to permit proper compaction. The rollers most frequently used for the compaction of emulsion mixes are pneumatic tire, vibratory smooth steel-wheel, and static steel-wheel rollers. With thicker lifts of dense-graded mix, padfoot rollers are sometimes used for breakdown rolling. Finish with a steel-wheel roller. The rolling equipment and how it is to be used will depend upon the type, properties, and layer thickness of the emulsion mix.

To prevent trapping moisture in thicker applications, it may be necessary to aerate the mix. The aeration process involves lightly scarifying and blading the surface of the previously compacted mixture to open voids, which allows moisture to escape. The aerated mix will be reshaped and compacted when sufficient moisture has evaporated. Mixtures that do not require aeration may be spread to the required thickness immediately after mixing.

Emulsion Central Plant Mix (Cold)

The central or stationary plant mixing of emulsions allows for the production of quality mixes due to more precise control of materials. This method of mixing provides the following advantages over hot mix asphalt:

- Economy—Lower fuel consumption, mobility, and low capital cost in equipment. This mixing method is ideally suited for projects in remote locations.
- Low Emissions—The lower temperatures used in central plant cold mixed production results in lower emissions at the plant and job site.
- Safety—Lower material temperatures improve worker safety at the plant and job site.

▶ Mixing Plants

The production of quality cold mixes requires a well-controlled operation. The setup for a central cold mix plant may vary depending on the type of mix. However, as a minimum, it is recommended the plant have:

- a pugmill mixer,
- emulsion storage,
- metered pumps for emulsion and water,
- one or more aggregate feeder bins, and
- belt conveyors.

The pugmill should allow for variation in the mixing time to ensure proper coating of the aggregate with emulsion. Batch-type pugmills can be used, but the production of these mixes is ideally suited for continuous type mixers (Figure 7.6).

☐ **Figure 7.6**   Cold Mix Continuous Plant

► Laydown and Compaction

For plant-produced emulsion mixes, the paving procedures are similar to those used for hot mix asphalt. Conventional asphalt paving machines are recommended, however, base courses may be laid with self-propelled base spreaders or even with a blade. Proper paver operation is critical to obtaining a uniform and smooth mat. This includes keeping the paver speed constant and the level of mix on the augers at a nearly constant depth. The screed should not be heated. If mix sticks to the paver screed or tearing of the mat occurs, a change in the grade or formulation of the emulsion, or the water content of the mixture, may eliminate the problem.

Plant-produced emulsion mixes are typically placed in lifts of 2 to 3 inches (50 to 75 millimeters) and can be placed in lifts of up to 4 inches (100 millimeters). For open-graded mixes, the breaking of the asphalt emulsion usually occurs fairly quickly. With open-graded mixes, rolling normally may begin immediately behind the paver. The emulsion in dense-graded mixes typically is slower to break and cure. Higher moisture contents may require a delay in the compaction of dense-graded mixes. The amount of delay is dependent on the lift thickness and the curing conditions such as air temperature, humidity, and wind.

Breakdown rolling of open-graded mixes are best accomplished with a static steel-wheel roller. Vibratory rolling of open-graded mixes is not recommended because fracture of aggregate and asphalt de-bonding may occur. A light "choke" or "blotter" of a coarse sand or fine crusher screenings should be applied after steel-wheel rolling to prevent pick up and damage by traffic. The sand or screenings is applied with a conventional self-propelled chip spreader or rotary spreader at the rate of 6 to 12 pounds/square yard (3.3 to 6.6 kilograms/square meter).

► Precautions

The following precautions should be taken with cold asphalt emulsion-aggregate mixes to assure good performance:

- Dense-graded mixes normally are resistant to water damage during construction. However, if rain occurs before the mixture is cured, traffic should be kept off or limited until the necessary compaction is accomplished.
- The water content should be no more than is required to disperse the emulsion, achieve adequate coating and mix workability.
- Over-mixing may cause the emulsion to break prematurely or strip from the aggregate.
- For faster curing, place the mix in several thin layers rather than in a single thick lift.
- Entrapped water and petroleum distillates can create problems if sealed too soon.
- If raveling occurs, the loose material should be lightly broomed as soon as possible to prevent further damage to the surface. If the raveling continues, the surface should be fog sealed with a light application of a slow-setting emulsion (SS or CSS) diluted with water. The rate of application and amount of water for dilution can be varied as required to prevent further raveling and a tacky surface and pickup by traffic.

Emulsion Central Plant Mix (Warm)

The production of warm plant-mix using asphalt emulsion as the binder is somewhat comparable to the production of hot mix asphalt using asphalt cement. Benefits of emulsion central plant warm mixing:

- Lower fuel consumption
- Lower emissions at the plant and job site
- Improved worker safety at the plant and job site
- The ability to coat large-size and open-graded aggregates with thicker films
- Reduced asphalt aging
- Increased reclaimed asphalt pavement (RAP) usage

▶ Materials

The aggregates should meet the quality requirements of ASTM, AASHTO, state or federal agency, and other acceptable specifications. The production and blending of aggregates for these mixes require the same degree of quality control as hot mix asphalt.

The asphalt emulsions normally used for warm emulsion mixes are CMS-2, CMS-2h, HFMS-2, HFMS-2h, or polymer-modified grades of these emulsions. There are other specialty emulsions specifically developed for warm mix applications.

▶ Mixing Plants

The mixing plant may be either a batch or drum mix plant. On batch plants, the pug-mill mixing chamber should be vented to allow for steam to escape. Also, the asphalt transfer system must allow turning off or reducing the heat on all lines, pumps, and jacketed asphalt weigh buckets. The temperature of the emulsion should never reach the boiling point of water [212°F (100°C)], and is usually kept below 185°F (85°C).

▶ Mixing Temperature

The minimum mixing temperature is based on the amount of coating of aggregate particles as determined by ASTM D 2489, *Test for Degree of Particle Coating of Bituminous-Aggregate Mixtures.* The minimum acceptable percentage of coated particles will vary with aggregate gradation, particle shape, surface texture, asphalt content, and use of the warm mixed material. Warm asphalt emulsion mixes typically are produced at a temperature of 150 to 260°F (66 to 127°C).

▶ Construction Methods

Warm emulsion mixes typically are spread with a conventional paving machine. In general, the same construction procedures are used for these mixtures as for hot mix asphalt. Some warm emulsion mixtures, produced with emulsions having very soft asphalt residues and a small quantity of solvent, may be stockpiled and placed cold.

For warm emulsion mixes, compaction should begin when the mix will support the roller without shoving. The breakdown rolling can be completed with a pneumatic or double drum (tandem) vibratory roller followed by finish rolling with a steel-wheel roller.

Maintenance Mixtures

Maintenance patching is often completed using stockpiled aggregate mixtures produced either cold or warm. Proper preparation of the area to be repaired is essential. The Asphalt Institute publication *Asphalt in Pavement Maintenance*, Manual Series No. 16 (MS-16), provides procedures for making repairs that, if properly followed, will ensure successful results. The asphalt emulsions recommended for this purpose are CMS-2, CMS-2s, CMS-2h, HFMS-2, HFMS-2s, and MS-2. Other nonstandard grades have also been used with success for immediate use or stockpiling maintenance mixtures. Aggregates should meet the quality requirements outlined previously for asphalt-aggregate mixtures. The recommended aggregate properties are given in Table 7.2.

These maintenance mixtures consist of two types—one for immediate use and the other for up to six months storage or stockpiling.

Immediate Use Maintenance Mixes

Asphalt emulsions can be used very effectively in the preparation of maintenance mixtures for immediate use. The emulsion-aggregate mixture can be mixed in a pugmill and transported to the job site.

Asphalt emulsion mixtures for immediate use may contain small amounts of solvents. The mixture does not gain full strength until the solvent evaporates. Patching mixes should not be produced with an excess amount of mixing water due to the curing time required before the patch can be opened to traffic.

Stockpile Maintenance Mixes

During the cold weather months, most maintenance mixes used are those stored in stockpiles. These mixes can be produced in late summer or early fall and stored in quantity for later use. Stockpile life depends considerably on the formulation of the asphalt emulsion. Mix workability comes from using an emulsion that contains some solvent. Stockpile life and workability at low temperatures are in direct proportion to the amount and type of solvent in the emulsion. The use of solvents in asphalt materials is increasingly being limited by air quality requirements. However, asphalt emulsions normally have less solvent content than the cutback asphalts used for stockpiled mixes.

A thin crust normally is formed on the surface of the stockpile, but beneath the crust the mix is still workable. The completed mixture should be stored in a clean area to prevent contamination and not stored in a low area or depression where water could get into the mix. If available, mix storage in a covered area provides the best protection and helps retain mixture workability.

Emulsion-Aggregate Mixes Using RAP

Reclaimed asphalt pavement (RAP) has many uses in pavement applications including the incorporation of RAP into emulsion-aggregate mixtures previously discussed in this chapter. The use of RAP in emulsion-aggregate mixtures typically lowers the design emulsion content. Handling, processing, and stockpiling are critical issues when using RAP. For more information, refer to Asphalt Institute MS-4, *The Asphalt Handbook*, Chapters 3 and 4.

CHAPTER 8

ASPHALT PAVEMENT RECYCLING

All types of asphalt pavements can be recycled: low, medium, and high traffic volume highways, county roads, city streets, airports, and parking lots. Recycling is defined as "the reuse, usually after some processing, of a material that already has served its first-intended purpose." Relative to asphalt pavement recycling, there are several methods available. Therefore, each project being considered for recycling must be carefully evaluated to determine the method most appropriate. The factors should include: (1) existing pavement condition, (2) existing pavement material types and thickness, (3) recycled pavement structural requirements, and (4) availability of recycling additives.

A candidate for recycling is any severely distressed and deteriorating roadway. The candidate may be a hot mix asphalt (HMA) (Figure 8.1) or surface treatment(s) on an aggregate base. The deteriorated condition may be due to the pavement being too thin or weak for the traffic. Poor drainage can also accelerate the pavement deterioration and should be addressed as part of the rehabilitation process.

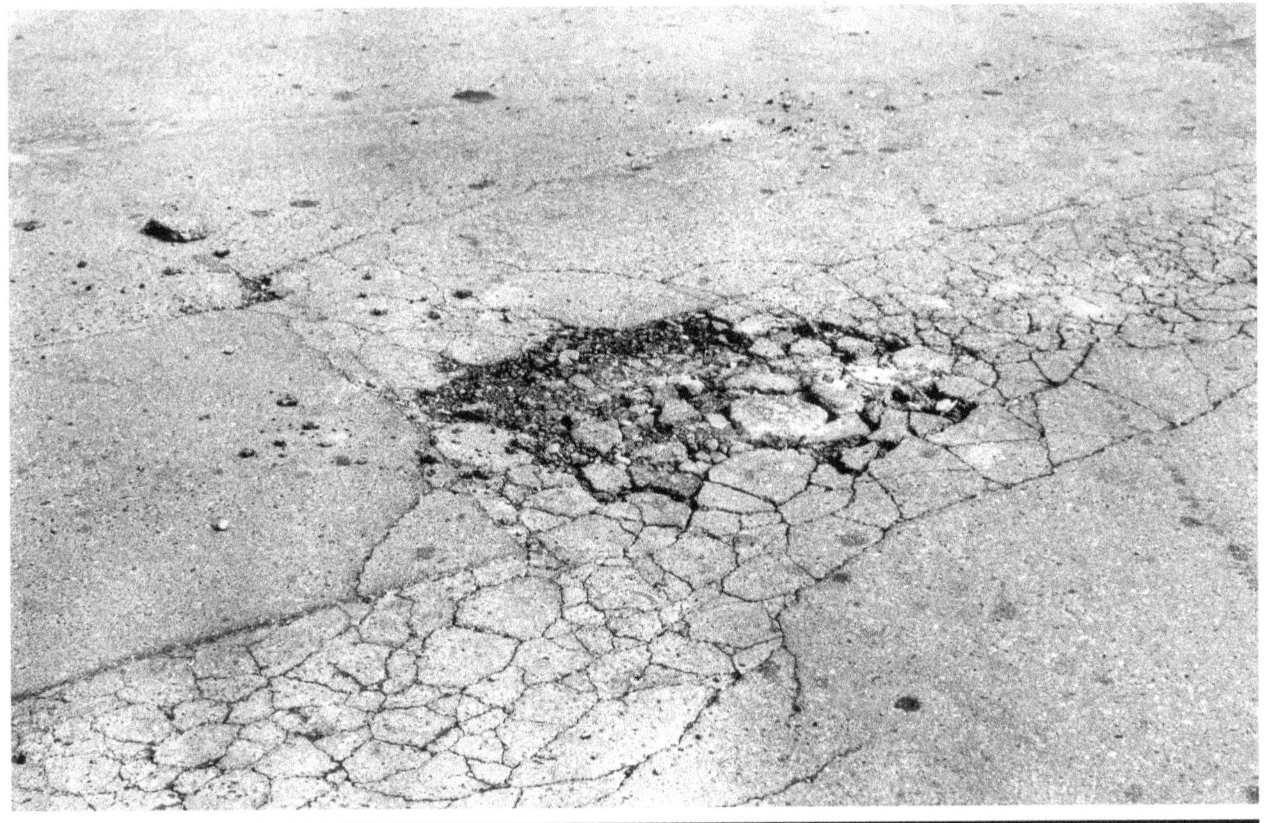

☐ **Figure 8.1**   Candidate for Recycling

Types of Recycling

The Asphalt Recycling and Reclaiming Association (ARRA) recognizes five types of asphalt pavement recycling:

Cold Planing

The asphalt pavement is removed to a specified depth, and the surface is restored to a desired grade and cross slope and free of humps, ruts, and other surface imperfections. The pavement removal or "milling" is completed with a self-propelled rotary-drum cold-planing machine. The reclaimed asphalt pavement (RAP) is transferred to trucks for removal and stockpiled for hot or cold recycling. Cold planed surfaces may be opened to traffic for short periods of time, provided that the pavement is monitored for raveling or other deterioration. Generally cold planed or milled surfaces require a new wearing course such as a new surface seal or HMA overlay. RAP generated by the cold planing process is an excellent source of recycling material as it needs very little, if any, processing to be re-used in either hot or cold recycling.

Hot Recycling

RAP is combined with new aggregate and asphalt cement and/or recycling agent in a hot mix asphalt plant to produce a recycled HMA. Both batch type and drum mix type hot mix plants are used to produce the recycled mix.

Hot In-Place Recycling

The recycling is performed on-site, in-place, and the pavement typically is processed to a depth of from 3/4 to 2 inches (20 to 50 millimeters). The asphalt pavement is heated, softened and scarified to the depth specified. An asphalt emulsion or other recycling agent is added to rejuvenate the age-hardened asphalt in the existing surface.

The three hot in-place recycling methods are:

- Surface recycling or heater-scarification
  - Process where the surface is heated, loosened, modified, remixed, and re-laid as a new wearing surface
    - Heating is done by radiant, infrared, or high velocity, super-heated hot air
    - Multiple rows of spring-loaded scarifiers or small milling heads are used to loosen the softened surface
    - Recycling agents are blended into emulsions which are then blended into the mixture
    - The mixture is then placed using conventional paving equipment.
- Remixing
  - Process where new HMA mixture is combined directly with the recycled mixture to significantly improve the recycling mixture before being placed
- Repaving
  - Process where the recycling surface is overlaid immediately with a new HMA mixture to form a thermal bond between the newly recycled mixture and the new HMA wearing surface.

Cold Recycling

Although cold recycling includes using the central or stationary plant process (see Chapter 7), the method discussed in this chapter is cold in-place recycling (CIPR). For CIPR, the existing asphalt pavement typically is processed to a depth of from 2 to 4 inches (50 to 100 millimeters) and the reclaimed material is mixed with an asphalt emulsion or emulsified recycling agent, spread and compacted to produce a new base course. Cold recycled bases have an open texture and require a new wearing surface to prevent water intrusion and raveling under traffic. Lower traffic pavements may use single or double chip seal surface treatment. Higher traffic pavements require a modified chip seal surface treatment or an HMA wearing surface.

Full-Depth Reclamation

With Full-Depth Reclamation (FDR), all of the pavement section, and in some cases a predetermined amount of underlying material, is mixed with asphalt emulsion to produce a stabilized base course. Base problems can be corrected with this construction. FDR consists of six basic steps: pulverization, additive and/or emulsion incorporation, spreading, shaping, compacting, and placement of a new wearing surface.

For more information on the processes mentioned above, refer to ARRA's *Basic Asphalt Recycling Manual* (BARM). The remainder of this chapter will focus on cold in-place recycling and full-depth reclamation, because these two methods predominately use emulsions and have become widely used.

An essential part of selecting any asphalt pavement rehabilitation method is determining the existing pavement condition. The type and amount of pavement defects need to be evaluated. The strength of the current pavement structure and its materials needs to be determined. The current and future traffic needs must be investigated.

Proper materials sampling and testing are very important to the success of pavement recycling. Pavement cores and/or test holes are used to determine the type, thickness, and condition of the various pavement layers and to obtain representative samples for laboratory testing. For in-place asphalt materials, testing typically includes conventional asphalt extraction for both asphalt content and aggregate sieve analysis. An asphalt recovery procedure should be included to determine the in-place asphalt properties. For aggregate bases and subgrade soils, a washed sieve analysis and sand equivalent or plasticity index are normally completed.

The typical tests performed for the CIPR and FDR processes are listed in Table 8.1. A mixture design is required to determine the type and amount of asphalt emulsion or emulsified recycling agent, premixing water content (if new aggregate is required), and the stability and strength properties of the recycled mixture.

◻ **Table 8.1: Materials Evaluation Procedures—
Cold In-Place Recycling and Full-Depth Reclamation**

Characteristics	Test Method	
	ASTM	AASHTO
Asphalt Content	D 2172	T 164
Asphalt Recovery	D 1856 or D 5404	T 170
Aggregate Sieve Analysis	C 136	T 27
Minus No. 200 (75 µm) Content	C 117	T 11
Liquid Limit, Plastic Limit, and Plasticity index	D 4318	T 89/90
Sand Equivalent Value	D 2419	T 176
Penetration	D 5	T 49
Viscosity	D 2171	T 202 or M 320

Advantages

Cold in-place recycling and full-depth reclamation of asphalt pavements provide the following advantages:

- Conserve resources by reuse/salvaging aggregate and asphalt in existing pavements.
- Pavement materials disposal is reduced or eliminated.
- Energy is conserved as the construction is completed in-place/on-grade.
- Reflective cracking is reduced or delayed with CIPR and eliminated by FDR.
- Improve or restore pavement crown and cross slope.
- Loss of curb reveal can be reduced or eliminated (Figure 8.2).
- Pavement maintenance costs are reduced.

☐ **Figure 8.2** Loss of Curb Height/Reveal and Drainage Capacity

Cold In-Place Recycling

The cold in-place recycling (CIPR) process includes milling the existing pavement to a prescribed depth [typically 2–4 inches (50–100 millimeters)] and blending with an asphalt emulsion as the stabilization agent. Depending on the process, the millings may be further processed through screening and crushing, and additional aggregate may also be incorporated into the mixture if needed. Processes used include a single-unit, a double-unit, or a multiple-unit train.

Single-Unit Train

The single-unit train consists of a milling or reclaiming machine that does the milling, RAP sizing, and blending at the cutting head. The maximum RAP size is a function of the condition of the existing pavement, speed of the cutting head, forward speed, and the machine's ability to retain large pieces in the milling chamber where they can be broken down. This process should not be used for pavements with severe alligator cracking as it is difficult to control the maximum particle size to less than 2 inches (50 millimeters). A computer-controlled spray bar located inside the cutting chamber adds the asphalt emulsion either from a tank on the milling machine or from a tanker. Aggregate or dry additives may be added to the RAP by spreading the material on the pavement ahead of the milling/reclaiming operation. The processed RAP is placed directly on the grade and placed by a separate paving operation, or it is deposited directly into a paver for immediate placement on the roadway (Figure 8.3).

☐ **Figure 8.3**   Single-Unit Train for Cold In-Place Recycling

Two-Unit Trains

The two-unit train consists of a milling/reclaimer and a pugmill mixer-paver (Figure 8.4). The pugmill mixer-paver is a cold mixing paver with an internal pugmill that is used to blend the RAP and asphalt emulsion. The two-unit train does not include any crushing and screening capability. The particle size is controlled in the same manner as in the single-unit train. The RAP is fed directly into a pugmill that has the capacity to accurately control the amount of asphalt emulsion and water being added to the RAP. After leaving the on-board mixer, the material is spread on the grade by the mixer-paver unit similarly to a conventional laydown paving machine.

○ **Figure 8.4**   Two-Unit Train for Cold In-Place Recycling

Multi-Unit Trains

A multi-unit train consists of a milling/reclaimer machine, a portable screening/crushing unit, and a pugmill mixer (Figure 8.5). The milling/reclaimer machine controls the depth of the pavement being milled. The RAP is picked up and deposited into the screening-and-crushing unit where the material is processed through a screen deck and any oversized material is returned to a crushing unit for resizing. The maximum size of the RAP is controlled by screens to a maximum particle size of 1–1 1/2 inches (25–37.5 millimeters).

○ **Figure 8.5**   Multiple-Unit Train for Cold In-Place Recycling

The weight of the RAP is monitored as it enters the pugmill where the asphalt emulsion is added and mixed. The amount of emulsion is controlled by a computerized metering system. The addition of a crushing-and-screening unit provides better mixture control and a more uniform CIPR recycling mixture than the single- and double-unit processes.

Placement of CIPR Materials

The CIPR mix is placed with conventional paving equipment. The mixture can be placed directly into the paver hopper for immediate placement, or it can be deposited in a window for placement within 24 hours. Unlike HMA paving, the screed should be operated cold to prevent the asphalt emulsion from breaking and sticking to the screed.

The emulsions typically used are the slow-setting and medium-setting grades: CMS, HFMS, CSS-1 and CSS-1h and polymer-modified versions of these grades. Additional aggregate can be added to improve gradation, stability, or cross-slope. Cement or lime, in dry or slurry form, can be added (1 to 1.5 percent by weight of RAP) to increase early strength and resistance to water damage.

Proper compaction procedures are required for CIPR mixes. With an uncompacted lift thickness up to 6 inches (150 millimeters), a delay in beginning the initial or breakdown rolling usually is required. The time of delay in rolling is dependent upon how rapidly the asphalt emulsion breaks, the depth of mix, and climatic conditions (temperature, humidity, and wind). With thicker lifts, a large pneumatic roller is usually specified as the breakdown (initial) roller. This roller is required to have a mass of 25 to 30 tons (23 to 27 tonnes) and tire pressures of 90 pounds/square inch (620 kiloPascal). A tandem vibratory steel-wheel roller normally is used for both intermediate and finish rolling. The finish rolling should be done in the static mode.

If raveling occurs, the amount of emulsion in the cold recycled mix may need to be increased in any further work. Additionally, a fog seal of diluted slow-setting emulsion may be applied to the compacted pavement to inhibit or control raveling. Adequate curing of the cold recycled mixes is required prior to placement of any new asphalt surface course. Inadequate curing can trap moisture which increases the possibility of asphalt stripping after the surface is placed.

Full-Depth Reclamation

When an asphalt pavement has insufficient structural strength, FDR with asphalt emulsion may be the solution. The depth of FDR depends upon the existing pavement thickness, subgrade soil conditions, and future traffic. The depth typically ranges from 6 to 10 inches (150 to 250 millimeters). The existing asphalt mix, granular base, and sometimes subgrade soil are processed on-grade and in-place, and treated with an asphalt emulsion to produce a new pavement with improved load-carrying capacity when fully cured.

The pulverization of the asphalt pavement and mixing of the asphalt emulsion normally are accomplished with reclaimer machines (Figure 8.6). During pulverization, new aggregate or additional RAP may be added by spreading on the grade ahead of the pulverization operation. The emulsion may be added through the reclaimer during pulverization or to a flattened windrow of the RAP material using a distributor. The emulsion is typically added through a metering system on the reclaiming machine, which provides better control of the asphalt content and increases productivity.

◻ **Figure 8.6** Reclaimer Machine

Depths greater than 4 inches (100 millimetres) may require vibratory padfoot rollers to achieve compaction. The padfoot rolled surface should be smoothed and shaped with a motor grader and finish rolled with a pneumatic-tire or steel-wheel roller. Compaction of FDR less than 4 inches (100 millimeters) may be accomplished using only a pneumatic-tire or steel-wheel roller.

When rolling results in cracking or severe displacement of the mix, it should be discontinued until the problem is identified and eliminated. The problem may be an underlying weak subgrade. This condition may be corrected by drying or stabilization with cement, fly ash, or lime.

An important step in the FDR process is the final grading of the surface. This work typically is completed with a motor grader. The final grading is particularly important if the final surface is to be an asphalt emulsion surface treatment. For some highway and airport projects, the finished surface may be established by trimming with a cold planing/milling machine having automatic grade and cross slope control and using a fixed reference or stringline.

As with other types of asphalt emulsion paving, FDR bases require waterproofing and a wearing course. Although HMA is generally used, a surface treatment may be all that is required for low traffic volumes.

CHAPTER 9

ADDITIONAL APPLICATIONS

The previous chapters in this manual have described the use of asphalt emulsions for mixes and various types of surface treatments. Asphalt emulsions can also be used for a number of other applications connected with both the construction and maintenance of paved surfaces. This chapter does not cover every possible use but offers guidelines for the more common ones.

Tack/Bond Coat

This application has traditionally been called a tack coat, but the trend is to use the term bond coat. A bond coat is a very light spray application of diluted asphalt emulsion (Figure 9.1). It is used to promote a bond between the existing surface and the new asphalt application. A bond coat is typically recommended for all overlays.

☐ **Figure 9.1** Applying a Bond Coat

The bond coat should be placed on a clean, dry pavement free of any materials that might prevent bonding. The application rate will vary depending on the type and condition of the surface being overlaid. The goal is to place a thin, uniform coating of emulsion covering 90–100 percent of the pavement surface. The application rate for a bond coat is normally 0.05–0.15 gallons per square yard (0.25–0.70 liter per square meter) of diluted slow setting cationic or anionic emulsion (SS-1, SS-1h, RS-1, CRS-1, CSS-1, CSS-1h, or CQS-1h). The emulsion is typically diluted at a rate of one part water to one part emulsion. If a given emulsion contains 65 percent residual asphalt in its original state, then after a 50-50 dilution, the product applied to the roadway would contain a 32.5 percent residual asphalt.

Spread rate and uniformity are difficult to control when using a distributor set at a low application rate. By diluting the bond coat, the emulsion can be applied at a higher spread rate while providing a uniformly thin coating of residual asphalt. A low application rate is usually applied between new or un-trafficked pavement layers. An intermediate rate is applied when resurfacing an existing, relatively smooth asphalt pavement. The highest rate is applied on old oxidized, cracked, pocked, or milled HMA or PCC pavement surfaces. Distributor operations should be monitored to prevent the accidental application of too much emulsion, which commonly occurs in those areas where the distributor starts and stops. If this happens, the distributor should be prevented from applying bond coat until the problem with the distributor or its operations is corrected. The excess material should be squeegeed over a larger area to thin out the excess to a more acceptable application rate.

The emulsion should break before placement of subsequent asphalt application. Shortly after placement of the bond coat, the emulsion will break. The color will change from brown to black, and the water will begin to evaporate. Once the water has evaporated, the physical properties of the residual asphalt will return and result in a strong adhesive bond. The rate of this breaking process will depend on the environmental conditions and the grade of emulsion selected. Generally, the emulsions used for bond coats will break and be fully cured shortly after placement. During this process, traffic and equipment (rollers, delivery trucks, etc.) should be kept off the surface.

Over-application of bond coats can form a slip plane. Excessive asphalt may eventually flush to the surface. Work should be planned so that no more than the necessary bond coat for the paving shift is placed on the surface. Vertical faces of existing pavements such as curbs and gutters should also be sprayed or painted with a uniform coating of approved bond material. This work should be done such that exposed curb or gutter surfaces are not stained. The use of low penetration tack/bond material such as an SS-1h or CSS-1h will also reduce the amount of tracking. Tracking is the unwanted coating or staining of adjacent pavement surfaces, both on and off the project, by trucks or equipment tires. Quicker breaking and trackless bond coats have been developed and a local supplier should be contacted for more information.

Prime Coat

Prime coats are spray applications of specialized asphalt emulsions on granular base in preparation for placing an asphalt mixture. These prime coat emulsions have been formulated specifically to penetrate an aggregate base. Properties for these prime coat emulsions have been established by local agencies. "Penetrating emulsion prime" (PEP) and "asphalt emulsion prime" (AEP) grades are examples of materials used for prime coats.

A prime coat performs several important functions, including:

- Coating and bonding loose mineral particles on the surface of the base.
- Hardening or toughening the surface of the base.
- Waterproofing the surface of the base by plugging capillary or interconnected voids.
- Providing adhesion or bond between the base and the asphalt mixture.

In order for the prime coat to satisfy these functions, some emulsion must penetrate into the base. The amount and size of voids and available moisture in the base material will influence emulsion penetration. When the base surface consists of fine grained materials, those passing the No. 200 (75 microns) sieve, the surface may act as a filter and inhibit the emulsion penetration. Dampening the surface and/or adding surfactants may overcome the filtering action. However, some surfaces may need to be loosened by scarifying prior to the emulsion being applied.

Crack Filling

Crack filling is an inexpensive, temporary preservation technique used to prevent the intrusion of water and debris into a crack. Crack filling is considered a lesser quality treatment than crack sealing. (Refer to *Asphalt in Pavement Maintenance*, (MS-16), Asphalt Institute for further information on crack sealing.) Crack filling is most effective in small, non-working cracks. Emulsions used in this application are typically applied using a cold pour pot and without any crack preparation. For the emulsions used for crack filling, consult a local emulsion manufacturer.

Patching

Emulsions are used in manufacturing cold mix patch materials and in the spray-injection patch process.

The methods typically used for patching using asphalt emulsion cold mixes are throw-and-roll, semi-permanent, and full-depth removal and replacement. All of these methods involve placing an emulsion mix into the repair area and compacting with a truck tire, vibratory plate compactor, or roller. Maintenance mixes for these repair methods and other patching mixes are covered in Chapter 7 of this manual. For information on pavement maintenance procedures, refer to *Asphalt in Pavement Maintenance*, (MS-16), Asphalt Institute.

An alternate method for repairing potholes or for skin patching is by spray injection. A special piece of equipment, either trailer- or truck-mounted, combines and blows asphalt emulsion and coarse, crushed aggregate into the repair area (see Figure 9.2). The spray-injection procedure consists of these steps:

- Blowing of water and debris from the repair area;
- Spraying a bond coat of asphalt emulsion in or on the repair area;
- Blowing of emulsion and aggregate in or on the repair area;
- Covering the repaired area with a thin layer of aggregate; and
- Opening the repair to traffic as soon as workers and equipment are clear.

This method of repair requires no compacting after the cover aggregate has been placed.

☐ **Figure 9.2** Spray-Injection Process

The asphalt emulsion used for spray injection varies between summer and winter application. Summer application, for temperatures above 50°F (10°C), works best with CRS-2, RS-2, or HFRS-2 grades. Limiting the penetration of the residue to a maximum of 135 has also shown to be beneficial in the performance of spray-injected patches placed in warm weather. Winter applications, [colder than 50°F (10°C)], call for a CMS-2, MS-2, or HFMS-2 emulsion. Requiring the penetration of the residue to be a minimum of 135 has also shown to be beneficial in the performance of spray-injected patches placed in cool weather.

For good aggregate coating, the emulsion temperature should be about 150°F (65°C), and the emulsion's Saybolt Furol viscosity at 122°F (50°C) should be limited to 250 seconds.

Aggregate sizes appropriate for spray injection are AASHTO or ASTM size No. 9 with no more than 3 percent passing the No. 200 (75 microns) sieve. Crushed aggregate material is recommended for spray injection. Using the emulsions described above, an emulsion content of approximately 5 percent by weight of aggregate is appropriate for warm weather conditions, while spray-injection patches placed in winter conditions perform well at an emulsion content of about 7 percent by weight of aggregate.

GLOSSARY

Aggregate

Aggregate: A hard inert mineral material, such as gravel, crushed rock, slag, or sand.

Coarse Aggregate: Aggregate retained on the No. 8 (2.36 millimeters) sieve.

Fine Aggregate: Aggregate passing the No. 8 (2.36 millimeters) sieve.

Sand: Fine aggregate resulting from natural disintegration and abrasion of rock or processing of completely friable sandstone.

Dense-Graded Aggregate: Aggregate that is graded from the maximum size down through filler with the object of obtaining an asphalt mix with a controlled void content and high stability.

Open-Graded Aggregate: Aggregate containing little or no mineral filler or in which the void spaces in the compacted aggregate are relatively large and interconnected.

Sandy Soil: A material consisting essentially of fine aggregate particles smaller than No. 8 (2.36 millimeters) sieve and usually containing material passing a No. 200 (75 microns) sieve. This material usually exhibits some plasticity characteristics.

Reclaimed Asphalt Pavement (RAP): Existing asphalt mixture that has been pulverized, usually by milling, and is used like an aggregate in the recycling of asphalt pavements.

Asphalt

Asphalt: "A dark brown to black cementitious material in which the predominating constituents are bituminous which occur in nature or are obtained in petroleum processing" (ASTM D8). Asphalt is a constituent in varying proportions of most crude petroleum.

Asphalt Cement: Asphalt that is refined to meet specifications for paving, roofing, industrial, and special purposes. Heat is required to make it fluid.

Asphalt Prime Coat: An application of asphalt primer to an absorbent surface. It is used to prepare an untreated base for an asphalt surface. The prime penetrates or is mixed into the surface of the base and plugs the voids, hardens the top, and helps bind it to the overlying asphalt course.

Asphalt Primer: A fluid asphalt of low viscosity (highly liquid) that penetrates into a non-bituminous surface upon application.

Asphalt Bond Coat (Tack): A very light application of asphalt emulsion diluted with water. It is used to ensure a good bond between the surface being paved and the overlying new course. Bond coats have historically been referred to as "tack" coats.

Cutback Asphalt: Asphalt cement that has been liquefied by blending with petroleum solvents.

Asphalt Emulsion: An emulsion of asphalt cement and water that contains a small amount of an emulsifying agent. Emulsified asphalt droplets may be of either the anionic (negative charge) or cationic (positive charge).

Asphalt Emulsion

Emulsifying Agent or Emulsifier: The chemical added to the water and asphalt that keeps the asphalt in stable suspension in the water. The emulsifier determines the charge of the emulsion and controls the breaking rate.

Breaking: The phenomenon when the asphalt and water in the emulsion separate, beginning the curing process. The rate of breaking is controlled primarily by the emulsifying agent.

Curing: The development of the mechanical properties of the asphalt binder. This occurs after the emulsion has broken and the emulsion particles coalesce and bond to the aggregate.

Residue: The asphalt binder that remains after the emulsion has broken and cured.

Equipment

Aggregate Spreaders: Machines used for spreading aggregate evenly at a uniform rate on a surface.

Mechanical Spreaders: Spreader boxes that are mounted on wheels. The spreaders are attached to and pushed by dump trucks.

Self-Propelled Spreaders: Spreaders having their own power units and two hoppers. The spreader pulls the truck as it dumps its load into the receiving hopper. Conveyor belts move the aggregate forward to the spreading hopper.

Tailgate Spreaders: Boxes with adjustable openings that are attached to and suspended from the tailgates of dump trucks.

Aggregate Trucks: Trucks equipped with hydraulic lifts to dump the aggregate into the spreader.

Asphalt Distributor: A truck or a trailer having an insulated tank and a heating system. The distributor applies asphalt to a surface evenly and at a uniform rate.

Cold In-place Recycling Train: A unit consisting of a large milling machine towing a screening/crushing plant and pugmill mixer for the addition of asphalt emulsion and production of cold mix base.

Milling Machine: A self-propelled unit having a cutting head equipped with carbide tipped tools for the pulverization and removal of layers of asphalt materials from pavements.

Reclaiming Machine: A self-propelled unit having a transverse cutting and mixing head inside a closed chamber for the pulverization and mixing of existing pavement materials with asphalt emulsion.

Asphalt emulsion (and mixing water) may be added directly through the machine by a liquid additive system and spray bar.

Rotary Power Broom: A power operated rotary broom used to clean loose material from the pavement surface.

Vacuum Broom: Usually a truck-mounted vacuum unit with a holding bin for debris. They also employ rotating broom heads to sweep debris into the path of the vacuum unit.

Spray-Injection Patching Machine: A special piece of self-contained equipment either truck- or trailer-mounted that mixes and blows an asphalt emulsion aggregate mix into a repair area. They have storage for emulsion and aggregate.

Pneumatic-Tire Rollers: Rollers with a number of tires spaced so their tracks overlap while applying kneading compaction.

Steel-Wheel Static Rollers: Tandem or three-wheel rollers with cylindrical steel rolls that apply their weight directly to the pavement.

Steel-Wheel Vibratory Rollers: A roller having single or double cylindrical steel rolls that apply compactive effort with weight and vibration. The amount of compactive force is adjusted by changing the frequency and amplitude of vibration.

Stationary Plants: Mixing plants located at a mixing site that mix asphalt emulsion and aggregates hot, warm, or cold. The simplest plants are for cold mixing and consist of aggregate feeder bin(s), asphalt and water metering systems, and a pugmill for mixing.

Equipment (continued)

Travel Plants: Self-propelled pugmill plants that proportion and mix aggregates and asphalt as they move along the road. There are three general types of travel plants:

1. One that moves through a prepared aggregate windrow on the roadbed, adds and mixes the asphalt as it goes, and rear discharges a mixed windrow ready for aeration and spreading.

2. One that receives aggregate into its hopper from haul trucks, adds and mixes asphalt, and spreads the mix to the rear as it moves along the roadbed.

3. Batch mixing units, such as slurry machines, that haul materials to the site and then mix and apply the materials

Types of Asphalt Surface Treatments and Mixes

Asphalt Emulsion Mix (Hot): A mixture of asphalt emulsion and mineral aggregate usually prepared in a conventional hot mix asphalt plant at a temperature less than 260°F (125°C). It is spread and compacted at a temperature above 200°F (95°C).

Pavement Base and Surface: The lower or underlying pavement course atop the subbase or subgrade and under the top or wearing course.

Plant Mix (Cold): A mixture of asphalt emulsion and mineral aggregate prepared in a central mixing plant and spread and compacted while the mixture is at or near ambient temperature.

Maintenance Mix: A mixture of asphalt emulsion and mineral aggregate for use in relatively small areas to patch holes, depressions, and distressed areas in existing pavements. Appropriate hand or mechanical methods are used in placing and compacting the mix.

Recycled Asphalt Mix: A mixture produced after processing existing asphalt pavement materials. The recycled mix may be produced by hot or cold mixing at a plant, or by processing the materials cold and in-place.

Asphalt Application: The application of sprayed asphalt coatings not involving the use of aggregates.

Asphalt-Aggregate Applications: Applications of asphalt material to a prepared aggregate base or pavement surface followed by the application of aggregate.

Single Surface Treatment: A single application of asphalt to a road surface followed immediately by a single layer of aggregate. The thickness of the treatment is about the same as the nominal maximum size aggregate particles.

Multiple Surface Treatment: Two or more surface treatments placed one on the other. The aggregate maximum size of each successive treatment is usually one-half the previous one, and the total thickness is about the same as the nominal maximum size of the first course. A multiple surface treatment may be a series of single treatments that produces a pavement course up to 1 inch (25 millimeters) or more in thickness. A multiple surface treatment is a denser wearing and waterproofing course than a single surface treatment, and it adds some strength.

Seal Coat: A thin surface treatment used to improve the surface texture and protect an asphalt surface. The main types of seal coats are surface treatments, fog seals, sand seals, slurry seals, micro surfacing, cape seals, and sandwich seals.

Fog Seal: A light application of diluted asphalt emulsion. It is used to renew old asphalt surfaces, seal small cracks and surface voids, and inhibit raveling.

Slurry Seal: A mixture of emulsified asphalt, well-graded fine aggregate, mineral filler or other additives, and water. It is applied from 1/8 to 3/8 inch (3 to 10 millimeters) thick and used to renew pavement surfaces and retard moisture and air intrusion into underlying pavement. Slurry seal will fill minor cracks, restore a uniform surface texture, and restore friction values.

Types of Asphalt Surface Treatments and Mixes

(continued)

Micro Surfacing: A mixture of polymer-modified asphalt emulsion, crushed dense-graded aggregate, mineral filler, additives, and water. Micro surfacing provides thin resurfacing of 3/8 to 3/4 inch (10 to 20 millimeters) to the pavement and returns traffic use in one hour under average conditions. Materials selection and mixture design make it possible for micro surfacing to be applied in multiple applications and provide minor reprofiling. The product can fill wheel ruts up to 1.5 inch (40 millimeters) in depth in one pass and produces high surface friction values. Micro surfacing is suitable for use on limited access, high-speed highways as well as residential streets, arterials, and roadways.

Cape Seal: A surface treatment where a chip seal is followed by the application of either slurry seal or micro surfacing.

Sandwich Seal: A surface treatment consisting of the application of a large aggregate, then a spray-applied asphalt emulsion (normally polymer modified), and covered with a smaller aggregate.

MISCELLANEOUS TABLES

☐ **Table B.1: Temperature-Volume Corrections for Asphalt Emulsion**

°C^t	°F	M*	°C^t	°F	M*	°C^t	°F	M*
10.0	50	1.00250	35.0	95	0.99125	60.0	140	0.98000
10.6	51	1.00225	35.6	96	0.99100	60.6	141	0.97975
11.1	52	1.00200	36.1	97	0.99075	61.1	142	0.97950
11.7	53	1.00175	36.7	98	0.99050	61.7	143	0.97925
12.2	54	1.00150	37.2	99	0.99025	62.2	144	0.97900
12.8	55	1.00125	37.8	100	0.99000	62.8	145	0.97875
13.3	56	1.00100	38.3	101	0.98975	63.3	146	0.97850
13.9	57	1.00075	38.9	102	0.98950	63.9	147	0.97825
14.4	58	1.00050	39.4	103	0.98925	64.4	148	0.97800
15.0	59	1.00025	40.0	104	0.98900	65.0	149	0.97775
15.6	60	1.00000	40.6	105	0.98875	65.6	150	0.97750
16.1	61	0.99975	41.1	106	0.98850	66.1	151	0.97725
16.7	62	0.99950	41.7	107	0.98825	66.7	152	0.97700
17.2	63	0.99925	42.2	108	0.98800	67.2	153	0.97675
17.8	64	0.99900	42.8	109	0.98775	67.8	154	0.97650
18.3	65	0.99875	43.3	110	0.98750	68.3	155	0.97625
18.9	66	0.99850	43.9	111	0.98725	68.9	156	0.97600
19.4	67	0.99825	44.4	112	0.98700	69.4	157	0.97575
20.0	68	0.99800	45.0	113	0.98675	70.0	158	0.97550
20.6	69	0.99775	45.6	114	0.98650	70.6	159	0.97525
21.1	70	0.99750	46.1	115	0.98625	71.1	160	0.97500
21.7	71	0.99725	46.7	116	0.98600	71.7	161	0.97475
22.2	72	0.99700	47.2	117	0.98575	72.2	162	0.97450
22.8	73	0.99675	47.8	118	0.98550	72.8	163	0.97425
23.3	74	0.99650	48.3	119	0.98525	73.3	164	0.97400
23.9	75	0.99625	48.9	120	0.98500	73.9	165	0.97375
24.4	76	0.99600	49.4	121	0.98475	74.4	166	0.97350
25.0	77	0.99575	50.0	122	0.98450	75.0	167	0.97325
25.6	78	0.99550	50.6	123	0.98425	75.6	168	0.97300
26.1	79	0.99525	51.1	124	0.98400	76.1	169	0.97275
26.7	80	0.99500	51.7	125	0.98375	76.7	170	0.97250
27.2	81	0.99475	52.2	126	0.98350	77.2	171	0.97225
27.8	82	0.99450	52.8	127	0.98325	77.8	172	0.97200
28.3	83	0.99425	53.3	128	0.98300	78.3	173	0.97175
28.9	84	0.99400	53.9	129	0.98275	78.9	174	0.97150
29.4	85	0.99375	54.4	130	0.98250	79.4	175	0.97125
30.0	86	0.99350	55.0	131	0.98225	80.0	176	0.97100
30.6	87	0.99325	55.6	132	0.98200	80.6	177	0.97075
31.1	88	0.99300	56.1	133	0.98175	81.1	178	0.97050
31.7	89	0.99275	56.7	134	0.98150	81.7	179	0.97025
32.2	90	0.99250	57.2	135	0.98125	82.2	180	0.97000
32.8	91	0.99225	57.8	136	0.98100	82.8	181	0.96975
33.3	92	0.99200	58.3	137	0.98075	83.3	182	0.96950
33.9	93	0.99175	58.9	138	0.98050	83.9	186	0.96925
34.4	94	0.99150	59.4	139	0.98025	84.4	184	0.96900
						85.0	185	0.96875

Legend: t = observed temperature in degrees Celsius (Fahrenheit)
   M = multiplier for correcting volumes to the basis at 15.6°C (60°F)
*Multiplier (M) for °C is a close approximation

Percent Depth Filled	Percent of Capacity	Percent Depth Filled	Percent of Capacity	Percent Depth Filled	Percent of Capacity	Percent Depth Filled	Percent of Capacity
1	0.20	26	20.73	51	51.27	76	81.50
2	0.50	27	21.86	52	52.55	77	82.60
3	0.90	28	23.00	53	53.81	78	83.68
4	1.34	29	24.07	54	55.08	79	84.74
5	1.87	30	25.31	55	56.34	80	85.77
6	2.45	31	26.48	56	57.60	81	86.77
7	3.07	32	27.66	57	58.86	82	87.76
8	3.74	33	28.84	58	60.11	83	88.73
9	4.45	34	30.03	59	61.36	84	89.68
10	5.20	35	31.19	60	62.61	85	90.60
11	5.98	36	32.44	61	63.86	86	91.50
12	6.90	37	33.66	62	65.10	87	92.36
13	7.64	38	34.90	63	66.34	88	93.20
14	8.50	39	36.14	64	67.56	89	94.02
15	9.40	40	37.39	65	68.81	90	94.80
16	10.32	41	38.64	66	69.97	91	95.55
17	11.27	42	39.89	67	71.16	92	96.26
18	12.24	43	41.14	68	72.34	93	96.93
19	13.23	44	42.40	69	73.52	94	97.55
20	14.23	45	43.66	70	74.69	95	98.13
21	15.26	46	44.92	71	75.93	96	98.66
22	16.32	47	46.19	72	77.00	97	99.10
23	17.40	48	47.45	73	78.14	98	99.50
24	18.50	49	48.73	74	79.27	99	99.80
25	19.61	50	50.00	75	80.39		

Printed in the USA
CPSIA information can be obtained
at www.ICGtesting.com
CBHW081817271024
16409CB00005B/17